A SIMPLIFIED APPROACH TO CONCRETE TECHNOLOGY

A Simplified Approach To CONCRETE TECHNOLOGY

(According to the current syllabus of West Bengal University of Technology)

By

P. K. NAG

Assistant Professor and Head, Department of Civil Engineering,
Institute of Engineering & Industrial Technology,
(A unit of Sanaka Educational Trust's Group of Institution)
Malandighi, Durgapur, Burdwan, (WB)

UNIVERSITY SCIENCE PRESS
(An Imprint of Laxmi Publications Pvt. Ltd.)
An ISO 9001:2008 Company

BENGALURU • CHENNAI • COCHIN • GUWAHATI • HYDERABAD
JALANDHAR • KOLKATA • LUCKNOW • MUMBAI • RANCHI • NEW DELHI
BOSTON (USA) • ACCRA (GHANA) • NAIROBI (KENYA)

A SIMPLIFIED APPROACH TO CONCRETE TECHNOLOGY

Printed and bound in India
Typeset at Kalyani Computer, New Delhi
First Edition: 2016
ISBN 978-93-5138-209-6

Branches

✆	Bengaluru	080-26 75 69 30	
✆	Chennai	044-24 34 47 26,	24 35 95 07
✆	Cochin	0484-237 70 04,	405 13 03
✆	Guwahati	0361-254 36 69,	251 38 81
✆	Hyderabad	040-27 55 53 83,	27 55 53 93
✆	Jalandhar	0181-222 12 72	
✆	Kolkata	033-22 27 43 84	
✆	Lucknow	0522-220 99 16	
✆	Mumbai	022-24 91 54 15,	24 92 78 69
✆	Ranchi	0651-220 44 64	

PUBLISHED IN INDIA BY

UNIVERSITY SCIENCE PRESS
(An Imprint of Laxmi Publications Pvt. Ltd.)

An ISO 9001:2008 Company
113, GOLDEN HOUSE, DARYAGANJ,
NEW DELHI - 110002, INDIA
Telephone : 91-11-4353 2500, 4353 2501
Fax : 91-11-2325 2572, 4353 2528
www.laxmipublications.com info@laxmipublications.com

C—
Printed at : Repro Knowledgecast Limited, Thane

To

My Beloved Parents

Late Sri Digendra Chandra Nag

and

Late Smt. Usha Rani Nag

Contents

Foreword

I have gone through the book entitled "A Simplified Approach to Concrete Technology" authored by my one time colleague Sri P. K. Nag, Department of Civil Engineering, Bengal College of Engineering & Technology, Durgapur. I find the same very useful to all undergraduate students of Civil Engineering. The book is well written in a lucid language with neat sketches and examples keeping in view the interest of average students.

Sri P. K. Nag has about 40 years of experience in one of the premier institutes named Central Mechanical Engineering Research Institute, Durgapur under Council of Scientific & Industrial Research, New Delhi, Prior to taking up teaching assignment, he had worked as a Consultant to a number of reputed organizations.

I hope this book will cater to the long-felt requirement of the West Bengal University of Technology students for a comprehensive book on Concrete Technology.

(N. P. Mukherjee)
Principal,
Bengal College of Engineering & Technology for Women,
Durgapur–713212.
[Formerly, Senior Scientist (Director's Grade),
Central Mechanical Engineering Research Institute, Durgapur.
Emeritus Scientist, CSIR, New Delhi
Ex-Principal, Dr. B. C. Roy Engineering College, Durgapur.]

Preface

This book is an outcome of my teaching experience over many years. Written primarily for the undergraduate students in civil engineering, it is proposed to present the art and science of concrete in a simple, clear, and scientific manner. The book completely covers the syllabus prescribed by West Bengal University and Technology and fulfils the dire need of suitable and simplified book on this topic. The main purpose of the book is to develop a sound understanding of the central concepts of concrete technology.

The approach of the average students of the B. Tech. courses to the study of this subject is found to be faltering. It has been felt that if this subject is explained in a very clear manner, it can attract the attention of the students and they can find it quite interesting.

It is a theoretical and practical subject and everybody is interested in studying it, at least from the point of view of passing the examination. As soon as a Civil Engineering graduate leaves the college and takes up a job, the most common work he is assigned to is to work on Concrete Technology; whetherhe is in the PWD, or the municipality, or the port trust, or the architect's office, because concrete is the most widely used building material.

Attempts have, therefore, been made to develop the subject from its very beginning in brief to avoid complications to the beginners. The large number of failures of structures has underlined the need for a better understanding of the behaviour of concrete. The use of concrete worldwide has increased the need to have a sound knowledge of the selection of materials, mix proportioning and quality control methods.

It may be added that due to its easy approach to the subject, this book will prove to be very helpful to the students studying for the degree courses in Civil Engineering. It is now for the students to say how far this humble attempt has been successful. Again, in a topic like "Mix Design of Concrete" the student has to appreciate the importance of writing correct procedure according to Bureau of Indian Standards (BIS). Emphasis has been given on Mix Design of Concrete for optimum use of concrete in construction.

It is felt that there are many scopes of improving the treatment of the subject and quite likely that some errors may be found due to hurry. It will be much appreciated if suggestions are made for improvements and errors are pointed out.

At the end, I pray to the Almighty that our Indian Civil Engineers would be able to build up a new India with efficiency, honesty and hard work.

—Author

Syllabus

WEST BENGAL UNIVERSITY OF TECHNOLOGY

(Civil Engineering Revised Course Structure)

SEMESTER V

Concrete as a Structural Material, Good Concrete Manufacture of Portland Cement, Chemical Composition of Cement, Hydration of Cement, Heat of Hydration and Strength, Tests on Cement and Cement Paste – fineness, consistency, setting time, soundness, strength.

Types of Portland Cement – Ordinary, Rapid hardening, low-heat, sulphate resisting, Portland slag, Portland pozzolana, super sulphated cement, white cement.

Aggregates – Classification, Mechanical and Physical Properties, Deletarious Substances, Alkali-Aggregate Reaction, Sieve Analysis, Grading Curves, Fineness modules, Grading Requirements. Testing of Aggregates – Flakiness, Elongation Tests, Aggregate Crushing Value, Ten Per cent Fines Value, Impact Value, Abrasian Value.

Quality of Water – Mixing Water, Curing Water, Harmful Contents.

Properties of Fresh Concrete – Workability, Factors Affecting Workability, Slump Test Compacting Factor Test, Kelly Ball Test, Flow Table Test, Segregation, Bleeding, Setting Time, Mixing and Vibration of Concrete, Mixers and Vibrators, curing, Methods, Maturity.

Strength of Concrete – Water/Cement ratio, Gel/Space ratio, Strength in Tension, Compression, Effect of Age on Strength, Relation between Compressive and Tensile Strength, Fatigue Strength, Stress Strain Relation and Modules of Elasticity, Poisson's Ratio, Shrinkage and Creep, Compression Test on Cubes, Cylinders, Non-Destructive Tests.

Admixtures – Different types, effects, uses, retarders and super plasticizers.

Mix Design by I.S. Code method.

Light-weight, Polymer and Fibre-reinforced concrete.

Chapter 1

CONCRETE

CONCRETE AS A STRUCTURAL MATERIAL

Concrete is used in structures like bridges or buildings, highways or dams. In order to use concrete satisfactorily, both the designer and the contractor need to be familiar with concrete technology. There are two commonly used structural materials, Concrete & steel. They sometimes complement one another. The man on the job needs to know more about concrete. It is possible to obtain concrete of specified quality. It is only the raw material that is bought transporting, placing and compacting greatly influence the final product. The selection of mixes cannot be made without a sound knowledge of the properties and behaviour of concrete.

What is Concrete?

Concrete is any product or mass made by the use of a **cementing medium**. This medium is the product of **reaction** between **hydraulic cement and water**. Now-a-days concrete is made with several types of cement and also containing pozzolan, fly ash, blast furnace slag, sulphur, admixtures, polymers, fibers, etc. A mixture of cement is of water, aggregate (fine and coarse) and admixtures. The use of aggregates in concrete is beneficial.

GOOD CONCRETE

Let us know first about a **bad concrete**. It is meant a substance with the consistence of soup, hardening into a honeycombed, non-homogeneous and weak mass. This material is made by mixing cement, aggregate and water.

We can make good concrete with the proper **Know–how**. There are two overall criteria. It has to be satisfactory in its **hardened state** and also in its **fresh state** while being transported from the mixer and placed in the formwork. The requirements in the fresh state are the consistence of the mix to be compacted at the job and the mix is **cohesive** enough to be transported and placed without **segregation**.

As far as the hardened state is considered, the usual requirement is a satisfactory compressive strength which is easy to measure. It is also ensured by water/cement ratio, cement content, density, etc.

MANUFACTURE OF PORTLAND CEMENT

Portland Cement is obtained by intimately mixing from a combination of a **calcareous material** such as **limestone or chalk and of silica and alumina** found as **clay or shale**. It is the **grinding of the raw materials** into a very fine powder, mixing them intimately and **burning** in a **large rotary kiln** at a temperature of about 1400°C (2550°F). The mass of the materials then **fuses** into **balls** known as **clinker**. The clinker is cooled and ground to a fine powder with some **gypsum** about 3 to 5% added and the resulting product is the commercial Portland Cement.

The mixing and grinding of the raw materials can be done either in water or in a dry condition which is known as **wet and dry process**. The Kiln, 7 m (23 ft) in diameter and 230 m (759 ft) long, is slightly inclined. The mixture is fed at the upper end while coal or other source of heat is blown in by an air blast at the lower end of the Kiln. The amount of coal required to manufacture one tone of Cement is between 100 kg to 350 kg depending on the process used.

FLOW CHART

Manufacturing of Cement

Limestone boulder from mines → **Clinkerization Section** → Crusher → Crushed Limestone → Conveyor Belt → Raw Meal → 4 stage Pre heater →

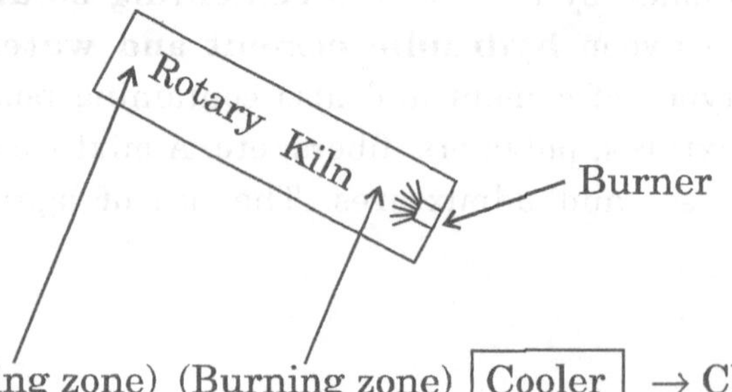

(Calcining zone) (Burning zone) Cooler → Clinker → **Cement Mill Section**

→ Clinker → Gypsum + Fly ash → Roller Press → Ball Mill → Cement Silo

→ Packing plant.

Chemical Composition of Cement

Raw materials → lime, silica, alumina and iron oxide.

Main Compounds in Portland Cement

Name of Compound	Oxide Composition	Formula
Tricalcium silicate	$3CaO\ SiO_2$	C_3S
Dicalcium silicate	$2CaO\ .\ SiO_2$	C_2S
Tricalcium aluminate	$3CaO\ .\ Al_2O_3$	C_3A
Tetracalcium aluminoferrite	$4CaO\ .\ Al_2O_3\ .\ Fe_2O_3$	C_4AF

Cement and Hydration of Portland Cement

Raw material for Cement — Limestone, Clay, Shale (Calcareous and argillaceous material)

↓

Component elements in raw materials — O Si Ca Al Fe

↓

Oxide Composition in raw materials — $CaO\ SiO_2\ Al_2O_3\ Fe_2O_3$

On burning Clinker formed

↓

Compound Composition — $C_3S\ C_2S\ C_3A\ C_4AF$

On grinding Clinker

↓

Portland Cement — Various types

On hydration

↓

Products of hydration — $C - S - H$ Gel + $Ca(OH)_2$

The Chemical reactions that take place between Cement and Water is referred as hydration of Cement.

HYDRATION OF CEMENT

As in Fig. 1.1, we know that the Cement is found in powder form but the material of interest in practice is the **set cement paste.** This is the product of reaction of Cement with water. In the presence of water, the **silicates (C_3S, C_2S) and aluminates (C_3A, C_4AF)** of Portland Cement from products of hydration or hydrates which in time produce a firm and hard mass—the hardened Cement Paste. The two Calcium silicates (C_3S and C_2S) are the main cementation compounds in Cement, the former hydrating much more rapidly than the latter. In commercial cements, the calcium silicates contain small impurities from some of the **oxides** present in the clinker. These impurities have a strong effect on the properties of the hydrated silicates. The 'Impure' C_3S is known as **alite** and the 'impure' C_2S as **belite**. The Calcium silicate hydrates are described as **C–S–H**. The approximate hydration reactions being written as follows:

For C_3S:

$$\underset{[100]}{2C_3S} + \underset{[24]}{6H} \rightarrow \underset{[75]}{C_3S_2H_3} + \underset{[49]}{3Ca(OH)_2}$$

For C_2S:

$$\underset{[100]}{2C_2S} + \underset{[21]}{4H} \rightarrow \underset{[99]}{C_3S_2H_3} + \underset{[22]}{Ca(OH)_2}$$

The numbers in the square brackets are the corresponding masses and on this basis both silicates require approximately the same amount of water for hydration but C_3S produces more than twice as more $Ca(OH)_2$ as is formed by the hydration of C_2S.

The amount C_3A in most cements is comparatively small. The reaction of C_3A with water is very rapid which is prevented by the addition of **gypsum** to the cement clinker.

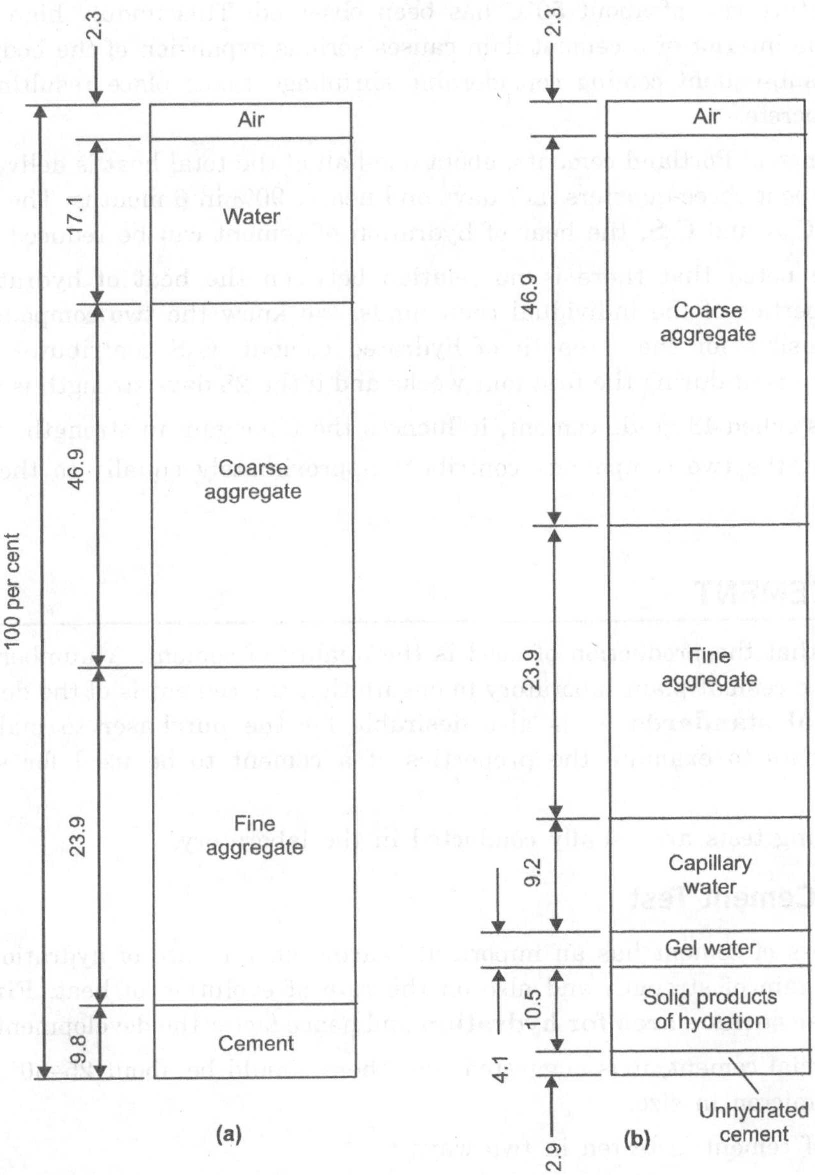

Fig. 1.1. Volumetric proportions of concrete of mix proportions 1 : 2 : 4 by mass with a water/cement ratio of 0.55 and entrapped air content of 2.3 per cent: (*a*) before hydration, and (*b*) when the degree of hydration is $h = 0.7$.

HEAT OF HYDRATION AND STRENGTH

The reaction of cement with **water** is exothermic. The reaction delivers a considerable quantity of **heat**. This can be easily observed if a cement with water is placed in a thermos flask. Much attention has been paid to the heat found during the hydration of cement in the interior of mass concrete dams. It is estimated that about 120 calories of heat is generated in the hydration of 1 gm of cement.

A temperature rise of about 50°C has been observed. This unduly high temperature developed at the interior of a cement dam causes serious expansion of the body of the dam and with the subsequent cooling considerable shrinkage takes place resulting in serious cracking of concrete.

For the range of Portland cements, about one-half of the total heat is delivered between 1 and 3 days, about three-quarters in 7 days and nearly 90% in 6 months. The reducing the proportions of C_3A and C_2S, the heat of hydration of cement can be reduced.

It may be noted that there is no relation between the heat of hydration and the cementing properties of the individual compounds. We know the two compounds C_3S and C_2S are responsible for the strength of hydrated cement. C_3S contributes most to the strength development during the first four weeks and if the 28 days strength is not less than 33 N/mm^2, it is called 43 grade cement, influences the later gain in strength. At the age of about one year, the two compounds contribute approximately equally to the strength of hydrated cement.

TESTS OF CEMENT

We know that the production of good is the quality of cement. A number of tests are performed in the cement plant laboratory to ensure that the cement is of the desired quality as per **national standards**. It is also desirable for the purchaser to make **periodic acceptance tests** to examine the properties of a cement to be used for some special purpose.

The following tests are usually conducted in the laboratory:

Fineness of Cement Test

The fineness of cement has an important bearing on the rate of hydration and hence on the rate of gain of strength and also on the rate of evolution of heat. **Finer cement offers a greater surface area for hydration** and hence faster the development of strength.

In commercial cement, it is suggested that there should be about 25–30% of particles of less than 7 micron in size.

Fineness of cement is tested in two ways :

(*a*) By sieving: **Sieve test is rarely used.**

(*b*) By determination of specific surface (**total surface area of all the particles in one gram of cement**) by air-permeability apparatus. Expressed as cm^2/gm or m^2/kg. Generally **Blaine air-permeability apparatus** is more commonly employed in India.

Consistence of Standard Paste Test

For finding out initial setting time, final setting time and soundness of cement and strength a parameter known as **standard consistency** or **normal consistency** has to be used. The standard consistency of a cement paste is defined as that consistency which will permit a **Vicat Plunger** having 10 mm diameter and 50 mm length to penetrate to a depth of 33–35 mm from the top of the mould. The Vicar Apparatus is used to find out the percentage of water required to produce a cement paste of standard consistency.

The following procedures is adopted to find out standard consistency. Take about 500 gms of cement and prepare a paste with a weighed quantity of water (say **24%** by weight of cement) for the **first trial**. The paste must be prepared in a standard manner and filled into the Vicat mould within 3–5 minutes. After completely filling the mould, shake the mould to expel air. A standard plunger, 10 mm dia. × 50 mm long is attached and brought down to touch the surface of the paste in the test block and quickly released allowing it to sink into the paste by its own weight. Take the reading by noting the depth of penetration of the plunger. Conduct a 2nd trial (Say with 25% of water). Similarly, depth of 33–35 mm from the top. That particular percentage of water which allows the plunger to penetrate only to a depth of 33–35 mm from the top is known as the percentage of water required to produce a cement paste of standard consistency with a temperature 27°C and humidity 90%.

Setting Time Test

Setting refers to a change from a **fluid** to a **rigid state**. It is the stiffening of the cement paste. Setting is mainly caused by a selective hydration of C_3A and C_3S and the temperature rises in the cement paste. There are two types of setting of the cement paste.

(*i*) Initial Setting Time (*ii*) Final Setting Time

The initial setting time is the time involving between the moment that the water is added to the cement to the time that the paste starts losing its plasticity. The final setting time is the time involving between the moment the water is added to the cement and the time when the paste has completely lost its plasticity and has attained sufficient firmness to resist certain definite pressure.

(i) Initial Setting Time

For the determination of initial set, the Vicat apparatus is used, this time with a 1 mm diameter needle, acting under a prescribed weight on a paste of standard consistence. We know that the period involving between the time when water is added to the cement and the time at which the needle penetrates the test block to a depth equal to 33–35 mm from the top is taken as initial setting time.

(ii) Final Setting Time

This time the Vicat apparatus is with a circular attachment with a centre needle. The cement shall be considered as finally set when lowering the attachment gently cover the surface of the test block, the centre needle makes an impression, while the circular cutting edge of the attachment fails to do so. The paste of the cement has attained such hardness that the centre needle does not go through the paste more than 0.5 mm. British Standards prescribe the initial setting time as a minimum time of 45 minutes and the final setting time as a maximum of 10 hours for Portland cements.

Soundness

It is very important that the cement after **setting** shall not undergo any appreciable change of volume. Certain cements have been found to undergo a large **expansion** after setting. Such expansion may occur due to reactions of **free lime, magnesia** and **calcium sulphate.** This will cause serious difficulties for the durability of structures when such cement is used.

The unsoundness in cement is due to the presence of **excess** of lime than that could be combined with **acidic oxide** at the Kiln. This is also due to inadequate burning or insufficiency in fineness of grinding. It is also likely that too high a proportion of **magnesium content** or **calcium sulphate content** may cause unsoundness in cement. For this reason the magnesia content allowed in cement is limited to 6%. The quantity of gypsum added will vary from 3 to 5% depending upon C_3A content. The excess of gypsum leads to an expansion in cement.

Unsoundness in cement is due to excess of lime, excess of magnesia or excessive proportion of sulphates. For testing purpose, it is determinated by **Le–Chatelier apparatus** for finding soundness of cement.

Strength

Cement–sand mortar and concrete of prescribed proportions are used for the purpose of determining the strength of cement. The **compressive strength of hardened cement** is the most important of all the properties. Strength tests are not made on neat cement. Three cubes are tested for compressive strength. It shall be the average of the strengths of the three cubes. Out of six sample cubes, three cubes shall be tested at **7 days** and remaining three cubes at **28 days**.

SUMMARY

- Concrete is used in structures like bridges or buildings, highways or dams.
- Concrete is any product or mass made by the use of a cementing medium. This medium is the product of reaction between hydraulic cement and water.
- Portland Cement is obtained by intimately mixing from a combination of a calcareous material such as limestone or chalk and of silica and alumina found as clay or shale.
- The mixing and grinding of the raw materials can be done either in water or in a dry condition which is known as wet and dry process.
- The reaction of cement with water is exothermic. The reaction delivers a considerable quantity of heat. This can be easily observed if a cement with water is placed in a thermos flask.
- A number of tests are performed in the cement plant laboratory to ensure that the cement is of the desired quality as per national standards.

- The fineness of cement has an important bearing on the rate of hydration and hence on the rate of gain of strength and also on the rate of evolution of heat.

REVIEW QUESTIONS

1. Why are tests on Cement necessary in a Cement Plant?
2. How can the heat of hydration of Cement be reduced?
3. Is there any relation between the Cementing properties and heat of hydration of Cement?
4. What are the causes of unsoundness of Cement?
5. Why is the C_3A content in Cement of interest? How does gypsum influence the hydration of C_3A?
6. Describe the effects of C_3S on the properties of Cement.
7. What is meant by the water of hydration of Cement?

Chapter 2

PORTLAND CEMENT

TYPES OF PORTLAND CEMENT

Portland Cement with different **chemical composition** may be exhibited different **properties**. It should thus be possible to select mixtures of raw materials for the production of cements with various desired properties. Several types of Portland Cement are available in the market and additional special cements can be produced for special uses.

The methods of manufacture have improved over the years. There has been a development of cements to serve different purposes with a change in specifications. Main types of Portland Cement are given below.

For general information, as per ASTM (American Society for Testing Materials) standards, cement is designated as Type I, Type II, Type III, Type IV, Type V and other minor types like Type IS, Type IP and Type IA, IIA and IIIA.

Ordinary Portland (Type I) Cement (OPC)

1. Ordinary Portland Cement 33 Grade - IS : 269–1989
2. Ordinary Portland Cement 43 Grade - IS : 8112–1989
3. Ordinary Portland Cement 53 Grade - IS : 12269–1987

It is the most common cement used in general concrete construction. It is an excellent general cement and is the cement most widely used. The OPC is classified into three grades, namely 33 grade, 43 grade and 53 grade depending upon the strength of the cement at 28 days. If the 28 days strength is not less than 33 N/mm^2, it is called 33 grade cement, if the strength is not less than 43 N/mm^2, it is called 43 grade cement and if the strength is not less than 53 N/mm^2, it is called 53 grade cement.

It has been possible to upgrade the qualities of cement by using high quality limestone, modern equipments, maintaining better particle size distribution, finer grinding and better packing. However, an excess of free lime causes unsoundness of the cement. In the characteristics of OPC, modern cements have a higher C_3S content and a greater fineness. OPC is grey in colour because of the presence of **iron oxide** in it.

Rapid-hardening Portland (Type III) Cement

As the name implies, the strength of this cement develops rapidly because of a higher C_3S content (up to 70%) and a higher fineness (minimum 325 m^2/kg). There is generally little difference in chemical composition between the OPC and the rapid-hardening Portland cements.

The principal reason for the use of Type III cement is when formwork is to be removed early for re-use or where sufficient strength for further construction is required quickly. This cement should not be used in mass concrete construction or in large structural sections because of its higher rate of heat development. The setting time of Type III and Type I cements is the same. The cost of Type III cement is only marginally greater than that of OPC.

Low-heat Portland (Type IV) Cement

This cement is generally used in large dams work. It has a low heat of hydration. There is a slower development of strength in low-heat Portland cement than with OPC because of the lower content of C_3S and C_3A but the ultimate strength is unaffected. The fineness must not be less than 320 m^2/kg to ensure strength.

Sulphate-resisting Portland (Type V) Cement

It is not a cement for general use. It should be specified only when necessary. This cement has a low C_3A content so as to avoid sulphate attack from outside the concrete. Otherwise the formation of Calcium Sulphoaluminate and gypsum (SO_3) would cause disruption of the concrete due to an increased volume of the resultant compounds.

To remedy the sulphate attack the use of cement with low C_3A content to 5% is found to be effective. Such cement with low C_3A and comparatively low C_4AF content is known as Sulphate Resisting Cement. This cement has a high silicate content (C_3S,C_2S). It is used in marine condition, in the construction of sewage treatment works, concrete is used for fabrication of pipes, etc.

Portland Slag (Type IS) Cement (PSC) (IS : 455–1989)

It is obtained by mixing Portland cement clinker, **gypsum** and granulated **blast furnace slag** which is a waste product in the manufacture of **Pig Iron**, in suitable proportions and grinding the mixture to get a thorough and intimate mixture between the constituents. It may also be manufactured by separately grinding Portland Cement **Clinker, gypsum** and ground granulated **blast furnace slag** and later mixing them intimately. The resultant product is a cement which has physical properties similar to those of ordinary Portland cement. It addition, it has low heat of hydration and is relatively better resistant to **chlorides,**

soils and **water** containing excessive amount of **sulphates** or **alkali** metals, **alumina** and **iron**, as well as, to **acidic waters** and therefore, this can be used for marine works. Sometimes, it is referred to as Portland blast–furnace cement.

Slag contains **lime, silica** and **alumina** but not in the same proportions as in Portland cement. The amount of slag should be between **25** and **70%** of the mass of the mixture. It is similar to ordinary Portland cement with respect to fineness, setting time, soundness and strength.

Portland Pozzolana (Types IP, P and I (PM)) Cements (PPC) (IS : 1489–1991)

Portland Pozzolana Cement (PPC) is manufactured by the inter-grinding of OPC clinker with 15 to 35% of pozzolanic material. A pozzolanic material is a siliceous or aluminous material which while in itself possessing no cementitious properties, which will, in finely divided form and in the pre-presence of water, react with calcium hydroxide at ordinary temperature, to form compounds prossessing cementitious properties. The pozzolanic materials generally used for manufacture of PPC are **calcined clay** or **fly ash**. **Fly ash is a waste material generated in the thermal power station when powdered coal is used as a fuel.**

It may be recalled that calcium silicates (C_3S, C_2S) produce considerable quantities of calcium hydroxide $\mathbf{Ca(OH)_2}$ which is by and large a useless material from the point of view of strength or durability. If such useless mass could be converted into a useful cementitious product, it considerably improves quality of concrete. The use of fly ash performs such a role. The pozzolanic action is shown below.

Calcium hydroxide + Pozzolana + Water PPC produces less heat of hydration. PPC consumes calcium hydroxide and does not produce calcium hydroxide as much as that of OPC. As the fly ash is finer and of lower density, the bulk volume of 50 kg bag is slightly more than OPC. PPC needs enough moisture for pozzolanic activity.

Portland Super Sulphated Cement (IS : 6909–1990)

It is manufactured by grinding together a mixture of 80–85% granulated slag, 10–15% hard burnt gypsum and about 5% Portland cement clinker. The product is ground finer than that of Portland cement. A fineness of 400 to 500 m²/kg is seen. It has a low heat of hydration. The advantages of this type of cement lie in a high resistance to sea water and to sulphate attack. The cement has to be stored under very dry conditions as otherwise it deteriorates rapidly.

Portland White Cement

For architectural purposes it is used. It is also used because of its low content of soluble **alkalis**. White cement is made from **china clay** which contains little **iron oxide** and **manganese oxide**, together with **chalk** or **limestone** free from specified impurities. The cost of white cement is high pastel colours can be obtained by adding pigments to the mixer, provided there is no effect on strength.

SOLVED QUESTIONS

Q. 1. How can the heat of hydration of Cement be reduced?

Sol. The heat of hydration depends on the chemical composition of the cement and is approximately equal to the sum of the heats of hydration of the individual pure compounds when their respective proportions by mass are hydrated separately as shown below.

Heat of hydration of pure compounds

Compound	Heat of hydration J/g	Heat of hydration Cal/g	Compound	Heat of hydration J/g	Heat of hydration Cal/g
C_3S	502	120	C_3A	867	207
C_2S	260	62	C_4AF	419	100

It follows that by reducing the proportions of C_3A and C_3S, the heat of hydration of cement can be reduced. Fineness of cement also affects the rate of heat development.

Q. 2. What are the raw materials used for the production of cement?

Sol. The typical raw materials used for making cement are limestone ($CaCO_3$), sand (SiO_2), stale clay (SiO_2, Al_2O_3 or Fe_2O_3) and iron ore (Fe_2O_3). Thus, the chemical components of cement are Calcium (Ca), Silicon (Si), Aluminium (Al) and iron (Fe). The Calcareous component lime (CaO) is derived from one of the following: limestone, chalk, marble, lime sand shell deposit, lime sludge. The argillaceous component (SiO_2, Al_2O_3 or Fe_2O_3) is derived from one of the following: clay, shale, tuff ash, slate, glass.

Raw materials used to derive the chemical components of cement

Calcium (Ca)	Silica (SiO_2)	Alumina (Al_2O_3)	Iron oxide (Fe_2O_3)
Limestone	Sand	Clay	Iron
Shale	Clay	Shale	Clay
Marl	Shale	Slag	Mill scale
	Marl		

Q. 3. What is clinker and how is it produced?

Sol. The main raw material for the production of Cement is clinker. Clinker is an artificial rock made by heating limestone and other raw materials in specific quantities to a very high temperature in a specially made Kiln.

Clinker is produced in a Kiln. A cement Kiln, a huge inclined rotating furnace, is a continuous stream process vessel. As the raw materials → limestone, clay and shale → tumble toward the 1500°C flame, the chemical reaction transforms them into clinker. Clinker production requires high temperatures which are generated by the combustion of coal. Alternative fuels can also be used in the Kiln.

Q. 4. What is the maximum size of fine aggregate?

Sol. We know that 4.75 mm size is the dividing line between the fine and coarse aggregate. Hence, 4.75 mm size is the maximum size of fine aggregate.

Q. 5. What are oversize and undersize of aggregates?

Sol. The particle size of aggregate retained on each sieve is the **oversize** (Coarser) than the sieve in question but **undersize** (Finer) than the sieve above.

Important notes

(*a*) **Pressure:**

- Psi = Pounds per square inch.
- MPa or N/mm^2 = Mega pascal = 10^6 Pascals.
- The unit of force is Newton (N) and that of pressure is Pascal (Pa). One Newton is the force required to accelerate 1 kg of mass by 1 m/sec^2. A stress of 1 N/m^2 is expressed as 1 Pa = $(1/10^6)$ N/mm^2.
- 1 kg = 9.81 N, say, 10 N

$$1 \text{ kN} = \frac{1 \times 1000}{10} = 100 \text{ kg}$$

- **For Cube Test:**

Load → 740 kN → 7,40,000 N

Area → 150 mm × 150 mm → 22,500 mm^2

$$\frac{740}{22.5} = 32.89 \text{ N/mm}^2$$

- **49 Tonnes** = $\dfrac{49000 \times 10}{150 \times 150} = \dfrac{4900}{225} = \dfrac{49}{2.25} = 21.78 \text{ N/mm}^2$

SUMMARY

- Portland Cement with different chemical composition may be exhibited different properties.
- Portland Pozzolana Cement (PPC) is manufactured by the inter grinding of OPC clinker with 15 to 35% of pozzolanic material.

REVIEW QUESTIONS

1. What are the main stages in the manufacture of Portland Cement?
2. What are the major compounds in Portland Cement?

Chapter 3

AGGREGATES

CLASSIFICATION

It can be classified as **(*i*) Normal weight aggregates, (*ii*) Light weight aggregates** and **(*iii*) Heavy weight aggregates.** Normal weight aggregates can be further classified as (*a*) Natural aggregates and (*b*) Artificial aggregates.

(a) Natural aggregates	**(b) Artificial aggregates**
(*i*) Sand, Gravel, Crushed	(*i*) Broken Brick
(*ii*) Rock such as Granite	(*ii*) Air-cooled Slag
(*iii*) Quartzite, Basalt	(*iii*) Sintered fly ash
(*iv*) Sandstone	(*iv*) Bloated clay

Aggregates can also be classified on the basis of the size of the aggregates as **(a) Coarse aggregate** and **(b) Fine aggregate.**

MECHANICAL AND PHYSICAL PROPERTIES

The mechanical and physical properties of normal weight aggregates with respect to **particle size, shape, surface texture, unit weight** and **porosity** are adopted with relation to the quality of aggregates. The aggregates generally **absorb water** and retain it in their **pores**. The shape of the particle may be **rounded, angular or irregular** depending on the type of the manufacturing process. The surface texture may be smooth with small pores or irregular with a large pock-marked surface. These are important parameters because they control the degree of water absorption and affect the workability of the concrete.

DELETERIOUS SUBSTANCES

There are three broad categories of deleterious substances that may be found in aggregate as noted below:

1. **Impurities** which interfere with the processes of hydration of Cement.
2. **Coatings** are preventing the development of good bond between aggregate and cement paste and
3. Certain individual particles which are **weak** or **unsound** in themselves. These harmful effects are distinct from those due to the development of chemical reactions between the aggregate and the cement paste. The aggregate may also certain sulphate or chloride salts.

ALKALI-AGGREGATE REACTION

Some of the aggregates contain reactive silica which reacts with alkalies present in cement *i.e.* sodium oxide and potassium oxide. It is proved that certain types of reactive aggregates are responsible for promoting alkali-aggregate reaction. The rate of deterioration may be slow or fast depending upon the conditions. Sandstones containing silica minerals quartz are found to be reactive. It is possible to reduce its tendency by altering either the proportion of reactive silica or its fineness. Alkali-aggregate reaction can be controlled by (*i*) Selection of non-reactive aggregates, (*ii*) By the use of low alkali cement, (*iii*) By controlling the void space in concrete, (*iv*) By controlling moisture condition and temperature.

SIEVE ANALYSIS

The process or operation of dividing a sample of aggregate into various fractions of same particle size is known as a sieve analysis. Its purpose is to determine the grading or gradation of the particle size distribution in a sample of aggregates. A Sample of air-dried aggregates are graded by shaking or vibrating the sieves with the largest sieve at the top. On completion of sieving the material retained on each sieve from the mesh shall be weighed.

The aggregates used for making concrete are normally of size 63 mm, 50 mm, 40 mm, 25 mm, 20 mm, 16 mm, 12.5 mm, 10 mm, 4.75 mm, 2.36 mm, 1.18 mm, 600 microns, 300 microns and 150 microns. The aggregate fraction from 63 mm to 4.75 mm are termed as coarse aggregate and those fraction from 4.75 mm to 150 microns are termed as fine aggregate. The size 4.75 mm is a common fraction appearing both in coarse aggregate and fine aggregate (C.A. and F. A.). Sieving can be done either manually or mechanically.

GRADING CURVES

The results of a sieve analysis can be reported in tabular form as shown below:

Example of sieve analysis

Sieve size (1)	Mass retained in gm (2)	Percentage retained (3)	Cumulative percentage passing (4)	Cumulative percentage retained (5)
10 mm	0	0	100	0
4.75 mm	6	2	98	2
2.36 mm	31	10.1	88	12
1.18 mm	30	9.8	78	22
600 μm	59	19.2	59	41
300 μm	107	34.9	24	76
150 μm	53	17.3	7	93
< 150 μm	21	6.8	—	—
	Total = 307			Total = 246
	Fineness modulus = 2.46			

Column (2) shows the mass retained on each sieve. Column (3) is the same quantity expressed as a % of the total mass of the sample. Column (4) shows the cumulative % (to the nearest 1%) passing each sieve and it is the % that is used in the plotting of the grading curve. In a grading chart where the ordinates represent the cumulative % passing and the abscissa are the sieve apertures plotted to a logarithmic scale.

Chart of a Grading Curve

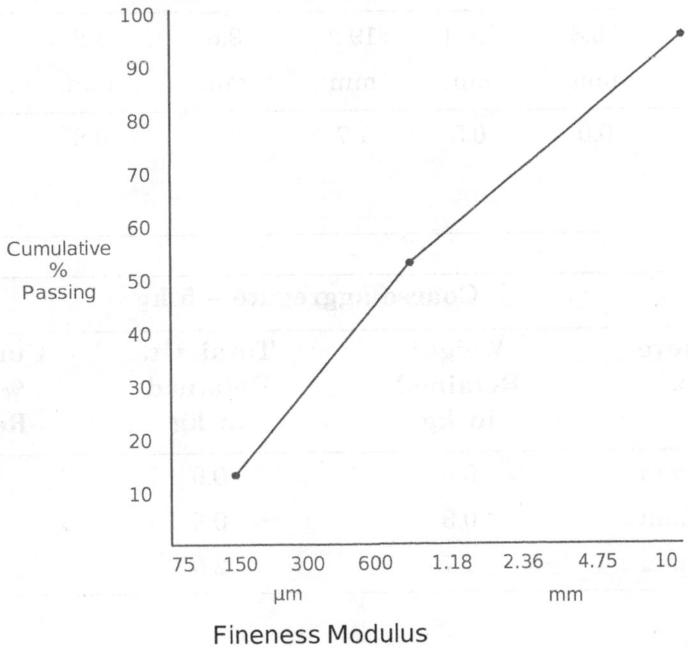

Solved Questions

Q. 1. Define the fineness modulus of aggregate.

Sol. **Fineness modulus** is a factor of coarseness or fineness of the material which is found from the sieve analysis the particle size distribution in a sample of aggregate. F.M. is an empirical factor obtained by adding the cumulative percentages of aggregate retained on each of the standard sieves ranging from 80 mm to 150 micron and dividing this sum by 100. The larger the figure, the coarser is the material. The following limits may be taken as guidance for fine aggregates.

Fine sand	:	Fineness Modulus	:	2.2–2.6
Medium sand	:	Fineness Modulus	:	2.6–2.9
Coarse sand	:	Fineness Modulus	:	2.9–3.2

A sand having a fineness modulus more than 3.2 will be unsuitable for making satisfactory concrete. A higher value is indicated a coarser grading.

Q. 2. What is meant by fineness modulus of aggregate?

Sol. **Coarse** and **fine** aggregates are sieved separately through the set of IS Sieves 76.8 mm, 38.4 mm, 1.92 mm, 9.6 mm, 4.8 mm and the set of IS Sieves No. 480, 240, 120, 60, 30 and 15 respectively. Sieves are placed in such a way that sieves of larger holes are on the top and smaller holes are on the bottom. The **cumulative percentage** of aggregate retained on each sieve is obtained. Find out the sum of cumulative percentages and divided it by 100 to get the fineness modulus of the aggregate.

Q. 3. Determine the fineness modulus of a 5 kg sample of coarse aggregate having the following sieve analysis:

IS Sieve No.	76.8 mm	38.4 mm	19.2 mm	9.6 mm	4.8 mm	240 no.
Weight Retained in kg.	0.0	0.3	1.7	2.1	0.8	0.1

Sol.

Coarse aggregate – 5 kg			
IS Sieve No.	**Weight Retained in kg**	**Total wt. Retained in kg**	**Cumulative % of wt. Retained**
76.8 mm	0.0	0.0	0.0
38.4 mm	0.3	0.3	6
19.2 mm	1.7	2.0	40

9.6 mm	2.1	4.1	82
4.8 mm	0.8	4.9	98
240 no.	0.1	5.0	100
120 no.	—	—	100
60 no.	—	—	100
30 no.	—	—	100
15 no.	—	—	100
	Total = 5.0		Total = 7.26

$$\text{Fineness modulus} = \frac{726}{100} \text{ or } 7.26\%$$

Q. 4. Is this aggregate suitable for good concrete?

Sol. We know for good concrete the value of **fineness modulus of coarse aggregate** should be between 6 to 8.5 and for **fine aggregate** between 2 to 3.5. Any increase in the value of **fineness modulus of aggregate** beyond the given range will result in a harsh mix, while decrease in the value will give an uneconomical mix.

In this problem the value of **fineness of aggregate** is 7.26 which is between 6 to 8.5. So it is suitable for good concreting.

GRADING REQUIREMENTS

We have seen how to find the grading of a sample of aggregate. Grading is of importance only in so far as it affects workability because strength is independent of the grading.

The Water and Cement ratio of the mix is generally fixed from strength or durability considerations. The amount of cement paste has to be sufficient to cover all of the particles so that the lower the surface area of the aggregate the less paste and therefore the less water is required. The workability is improved when there is an excess of paste above that required to fill the voids in the sand and also an excess of mortar (sand plus cement) above that required to fill the voids in the coarse aggregate because the fine material lubricates the larger particles. Another factor is the tendency of concrete to segregate for particles of different sizes to pack with smaller particles filling into the voids between larger ones.

TESTING OF AGGREGATES

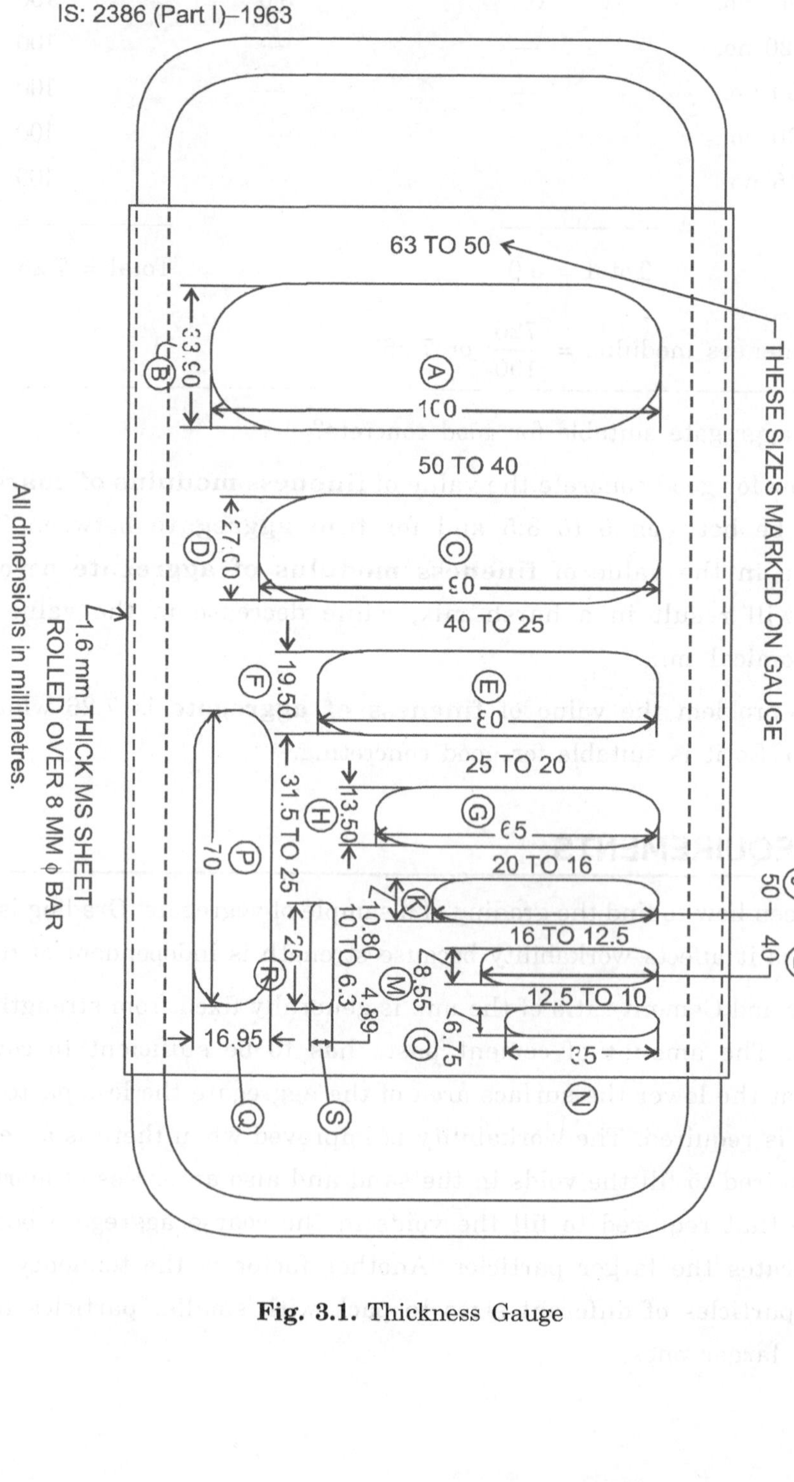

Fig. 3.1. Thickness Gauge

Flakiness

This method of test lays down for determining the **flakiness index** of coarse aggregate as per **IS : 2386 (Part I) – 1963.** The flakiness index of an aggregate is the percentage by weight of particles in it whose least dimension (thickness) is less than three-fifths of their mean dimension. The test is not applicable to sizes smaller than 6.3 mm. This test is conducted by using a metal thickness gauge as shown in Fig. 3.1. A sufficient quantity of aggregate is taken such that a minimum number of 200 pieces of any fraction to be tested. Each fraction is gauged in turn for thickness on a metal gauge of the pattern shown in Fig. 3.1. The total amount passing in the gauge is weighed to an accuracy of 0.1% of the weight of the samples taken. The appropriate size of material is shown in the standard dimensions specified in Column (3) of Table 3.1.

Table 3.1: Dimensions of thickness and length gauges

Size of Aggregate		Thickness Gauge* (mm)	Length Gauge** (mm)
Passing through IS Sieve (mm)	Retained on IS Sieve (mm)		
(1)	(2)	(3)	(4)
63	50	33.90	—
50	40	27.00	81.0
40	25	19.50	58.5
25	20	13.50	40.5
20	16	10.80	32.4
16	12.5	8.55	25.6
12.5	10	6.75	20.2
10	6.3	4.89	14.7

* This dimension is equal to 0.6 times the mean sieve size.

* * This dimension is equal to 1.8 times the mean sieve size.

The result is that the flakiness index is the total weight of the material passing the various thickness gauges or sieves expressed as a percentage of the total weight of the sample gauged.

Test for Determination of Elongation Index

This test is as per **IS : 2386 (Part-I)–1963**. The Elongation index on an aggregate is the percentage by weight of particles whose greatest dimension (length) is greater than 1.8 times their mean dimension. The elongation index is not applicable to sizes smaller than

6.3 mm. This is conducted by using metal length gauge as shown in Fig. 3.2. A sufficient quantity of aggregate is taken to provide a minimum number of 200 pieces of any fraction to be tested. Each fraction shall be gauged individually for length on the metal gauge. The gauge length used shall be that specified in Column (4) of Table 3.1 for the appropriate size of material. The total amount retained by the gauge length shall be weighed to an accuracy of at least 0.1% of the weight of the test samples taken. The elongation index is the total weight of the material retained on the various length gauges expressed as a percentage of the total weight of the sample gauged.

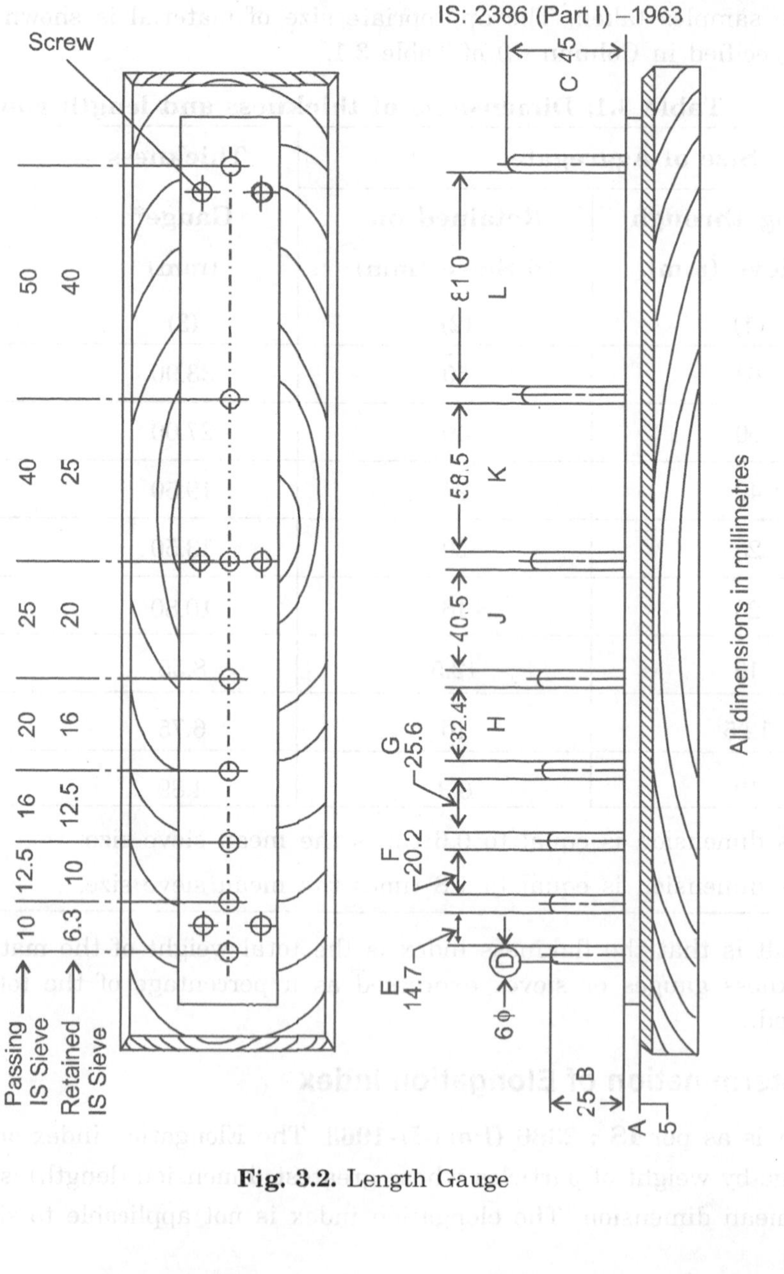

Fig. 3.2. Length Gauge

Test for Determination of Aggregate Crushing Value

(Mechanical Properties of Aggregates) IS : 2386 (Part-IV)-1963

The '**aggregate crushing value**' gives a relative measure of the resistance of an aggregate to crushing under a gradually applied compressive load. With aggregate of '**aggregate crushing value**' 30 or higher, the result may be anomalous and in such cases the '**Ten Percent Fines Value**' should be determined and used instead.

The standard aggregate crushing test shall be made on aggregate passing a 12.5 mm IS Sieve and retained on a 10 mm IS Sieve. If required or if the standard size is not available, other sizes upto 25 mm may be tested. But owing to the non-homogeneity of aggregates the results will not be comparable with those obtained in the standard test.

About 6.5 kg material consisting of aggregates passing 12.5 mm and retained on 10 mm sieve is taken. The aggregate in a surface dry condition is filled into the standard cylindrical measure in three layers approximately of equal depth. Each layer is tamped 25 times with the tamping rod and finally leveled off. The weight of the sample contained in the cylinder measure (A) is taken. The same weight of the sample is taken for the repeat test.

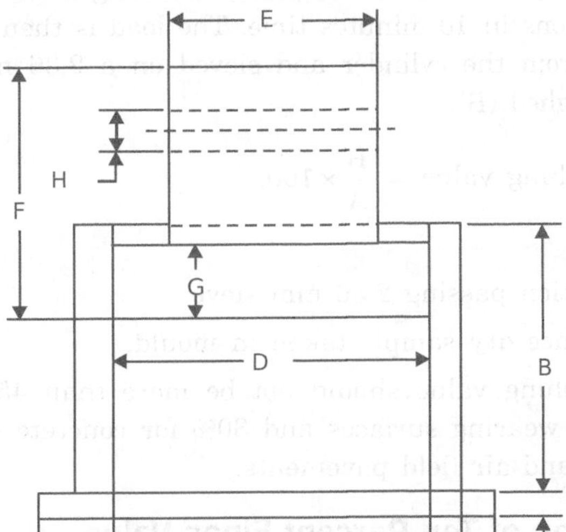

Key to Dimensions			
Letter	**Dimensions**	**150 mm Cylinder**	**75 mm Cylinder**
	Cylinder	**mm**	**mm**
A	Internal diameter	152.0 ± 0.5	77.0 ± 0.5
B	Height	130 to 140	70 to 80

C	Wall thickness	$\nless 16$	$\nless 8$
	Plunger		
D	Diameter of piston	150 ± 0.5	75.0 ± 0.5
E	Diameter of stem	100 to 150	50 to 75
F	Height	100 to 115	65 to 75
G	Depth of piston	$\nless 25$	$\nless 20$
H	Diameter of hole (nominal)	20	10
	Base-Plate		
J	Thickness (nominal)	6.3	6.3
K	Side length of equant	200 to 230	110 to 115

Fig. 3.3. Principal Dimensions of Apparatus for Aggregate Crushing Test

The cylinder of the test apparatus with aggregate filled, is put on the base plate and the plunger inserted horizontally on this surface as shown in Fig. 3.3. The plunger should not jam in the cylinder. It is then placed on the compression testing machine and is loaded uniformly upto a total load of 40 tons in 10 minutes time. The load is then released and the whole of the material removed from the cylinder and sieved on a 2.36 mm IS Sieve. The fraction passing the sieve is weighed (B).

$$\text{The aggregate crushing value} = \frac{B}{A} \times 100.$$

where,

B = Weight of fraction passing 2.36 mm sieve.

A = Weight of surface dry sample taken in mould.

The aggregate crushing value should not be more than 45% for aggregate used for concrete other than for wearing surfaces and 30% for concrete used for wearing surfaces such a runways, roads and air field pavements.

Test for Determination of Ten Percent Fines Value

As per IS : 2386 (Part-IV)–1963

This sample of aggregate for this test is the same as that of the sample used for aggregate crushing value test. The test sample is prepared in the same way. The cylinder of the test apparatus is placed in position on the base plate and the test sample added in thirds.

The apparatus with the sample and plunger is placed in the compression testing machine. The load is applied to a total penetration of the plunger in 10 minutes of about:

(*a*) 15 mm for rounded aggregates (say, gravels)

(*b*) 20 mm for normal crushed aggregates and

(*c*) 24 mm for honeycombed aggregates (say, slags).

After reaching the required penetration, the load is released and the whole material removed is sieved on a 2.36 mm IS Sieve. The fines passing the sieve is weighed and the weight is expressed as a percentage of the weight of the sample. The percentage would fall within the range 7.5 to 12.6 but if it does not, a further test shall be made at a load to bring the percentage fines within the range of 7.5 to 12.5%.

$$\text{Load required for 10\% fines} = \frac{14 \times X}{Y + 4}$$

where,

X = Load in tons causing 7.5 to 12.5%

Y = Mean percentage fines from two tests at X tones load.

Impact Value

The aggregate impact gives relative measure of the resistance of an aggregate to sudden shock or impact. The sample of aggregate for this test is the same as that of the sample used for aggregate crushing value test. The whole sample is filled into a cylindrical steel cup firmly fixed on the base of the machine. The net weight of the aggregate in the measure is determined (Weight A).

A hammer weighing about 14 kgs is raised to a height of 380 mm about the upper surface of the aggregate in the cup and allowed to fall freely on the aggregate. The test sample shall be subjected to a total 15 such blows each being delivered at an interval of not less than one second. The crushed aggregate is removed from the cup and the whole of it is sieved on 2.36 mm IS Sieve. The fraction passing the sieve is weighed (Weight B). The fraction retained on the sieve is also weighed (Weight C). If the total weight (B + C) is less than the initial weight A by more than one gm the result shall be discarded and a fresh test made. Two tests are made. The ratio of the weight of fines formed to the total sample weight in each test is expressed as percentage.

Therefore,

$$\text{Aggregate Impact Value} = \frac{B}{A} \times 100$$

where,

B = Weight of fraction passing 2.36 mm IS Sieve.

A = Weight of oven – dried sample.

The aggregate impact value should not be more than 45% for concrete other than wearing surfaces and 30% for concrete to be used as wearing surface, such as runways, roads and pavements.

Test for Determination of Aggregate Abrasion Value

This test covers the following two methods of determining the abrasion value of coarse aggregate as per **IS : 2386 (Part IV)–1963.**

(*a*) **By the use of Deval machine** and

(*b*) **By the use of Los Angeles machine**—gives a better realistic picture of the abrasion resistance of the aggregate.

The Los Angeles abrasion testing machine is shown in Fig. 3.2. The machine shall consist of a hollow steel cylinder, closed at both ends, having an inside diameter of 700 mm and an inside length of 500 mm.

The detail of **abrasive charge** which consists of Cast iron spheres or steel spheres approx. 48 mm in diameter and each weighing between 390 to 445 gm shall be as follows:

Grading	Number of spheres	Weight of charge (gm)
A	12	5000 ± 25
B	11	4584 ± 25
C	8	3330 ± 20
D	6	2500 ± 15
E	12	5000 ± 25
F	12	5000 ± 25
G	12	5000 ± 25

The test sample of aggregate is dried at 105° C to 110° C and it should conform to one of the gradings as shown in Table 3.2.

Table 3.2

Sieve size		Weight in gm of test sample for grade						
Passing mm	Retained on mm	A	B	C	D	E	F	G
80	63					2500		
63	50					2500		
50	40					5000	5000	
40	25	1250					5000	5000
25	20	1250						5000
20	12.5	1250	2500					
12.5	10	1250	2500					
10	6.3			2500				
6.3	4.75			2500				
4.75	2.36				5000			

Test sample and abrasive charge are placed in Los Angeles Abrasion testing machine and the machine is rotated at a speed of 20 to 33 rev/min. For gradings A, B, C and D, the machine is rotated for 500 revolutions. For gradings E, F and G, it is rotated 1000 revolutions. At the completion of the above number of revolution, the material is discharged from the machine. The material coarser than 1.7 mm IS Sieved is washed, dried in an oven at 105° to 110° C and weighed to the nearest gram.

The difference between the original weight and the final weight of the test sample is expressed as a percentage of the original weight of the test sample. This value is the percentage of wear. The percentage of wear should not be more than 16% for concrete aggregates.

QUALITY OF WATER

When we consider the **strength of concrete**, the quality of water in the mix may be affected the **hardening of concrete.** The quality of the water is important because **impurities** in it may interfere with the **setting** of the cement. It may also lead to **corrosion** of the reinforcement. For these reasons, the suitability of water for mixing and curing purposes should be considered. Clear distinction must be made between the effects of mixing water and the attack on hardened concrete by the water because it may be harmful when used in mixing.

Mixing Water

The suitability of water for mixing concrete is that if water is fit for drinking it is fit for making concrete. This does not appear to be a true statement for all conditions. Some waters containing a small amount of sugar would be suitable for drinking but not for mixing concrete and similarly water suitable for making concrete may not be fit for drinking. The drinking water may be unsuitable as mixing water when the water has a high concentration of **sodium** or **potassium** and there is a **danger of alkali–aggregate** reaction. The water may be accepted for making concrete if the **pH value** of water lies between 6 and 8 and the water is free from organic matter. If the compressive strength of concrete is found upto 90% from cube test, the source of water may be accepted. We should know about the harmful impurities in water. Salts of **Manganese, Tin, Zinc, Copper, Lead** and **Sodium sulphide** cause a reduction in strength of concrete. **sea water** also react with reactive aggregates as **alkalies** in cement.

Curing Water

Generally satisfactory water for mixing is also suitable for curing purposes. However, iron or organic matter may cause harmful for curing water. Any water suitable for mixing or even slightly inferior in quality, is acceptable for curing. However, the curing water should be free from the substances that attack hardened concrete. Curing with sea water may lead to attack of reinforcement.

SOLVED QUESTIONS

Q. 1. What are the requirements for water to be used for curing concrete?

Sol. It is in (*b*) Curing water.

Q. 2. Is drinking water always suitable as mix water?

Sol. It is in (*a*) Mixing water.

Q. 3. What are the dangers of using sea water as mixing water?

Sol. Sea water has a salinity of about 3.5%. In that about 78% is sodium chloride and 15% is chloride and sulphate of magnesium. Sea water also contain small quantities of sodium and potassium salts. This can react with reactive aggregates as alkalies in cement. The corrosion of reinforcement in concrete is found due to the use of sea water.

SUMMARY

- The mechanical and physical properties of normal weight aggregates with respect to particle size, shape, surface texture, unit weight and porosity are adopted with relation to the quality of aggregates.
- Fineness modulus is a factor of coarseness or fineness of the material which is found from the sieve analysis the particle size distribution in a sample of aggregate.
- The 'aggregate crushing value' gives a relative measure of the resistance of an aggregate to crushing under a gradually applied compressive load.
- The suitability of water for mixing concrete is that if water is fit for drinking it is fit for making concrete.
- Generally satisfactory water for mixing is also suitable for curing purposes. However, iron or organic matter may cause harmful for curing water.

REVIEW QUESTIONS

1. What is mechanical and Physical properties?
2. What is sieve analysis?
3. Discuss the quality of water.

Chapter 4

Properties of Fresh Concrete

WORKABILITY

Fresh concrete should be easy to mix, compact and finish as well as it must remain free from **segregation** during operations which may be termed as 'workability'. Slump test is carried to have a rough estimate of the workability of the concrete.

The quantity of water used in mixing is most important in determining the workability, strength, etc., of the concrete. An excess of water gives a weaker concrete. The ratio of the amount of water to the amount of cement by weight is known as 'water cement ratio'.

FACTORS AFFECTING WORKABILITY

The workability depends on a number of interacting factors: Water content, Aggregate type and Grading, Aggregate/Cement ratio and Fineness of cement. For a constant water and cement ratio, the workability increases as the aggregate/cement ratio is reduced because the amount of water relative to the total surface of solids is increased.

There are two other factors which effect workability: **time** and **temperature**. A higher temperature reduces the workability. Finer particles require more water to wet their larger specific surface. To achieve for minimum voids or for maximum density with no segregation, the influence of the aggregate type and grading has to be considered.

SLUMP TEST

Slump test, as shown in Fig. 4.1, is the most commonly used method of measuring consistency of concrete. It can be used either in laboratory or at site of work. The metallic mould for the slump test is a frustum of a cone, bottom diameter is 20 cm, top diameter is 10 cm and height is 30 cm. All are internal dimensions as stated above. The container is filled with concrete in three layers. Each layer is tamped 25 times with a standard 16 mm diameter steel rod.

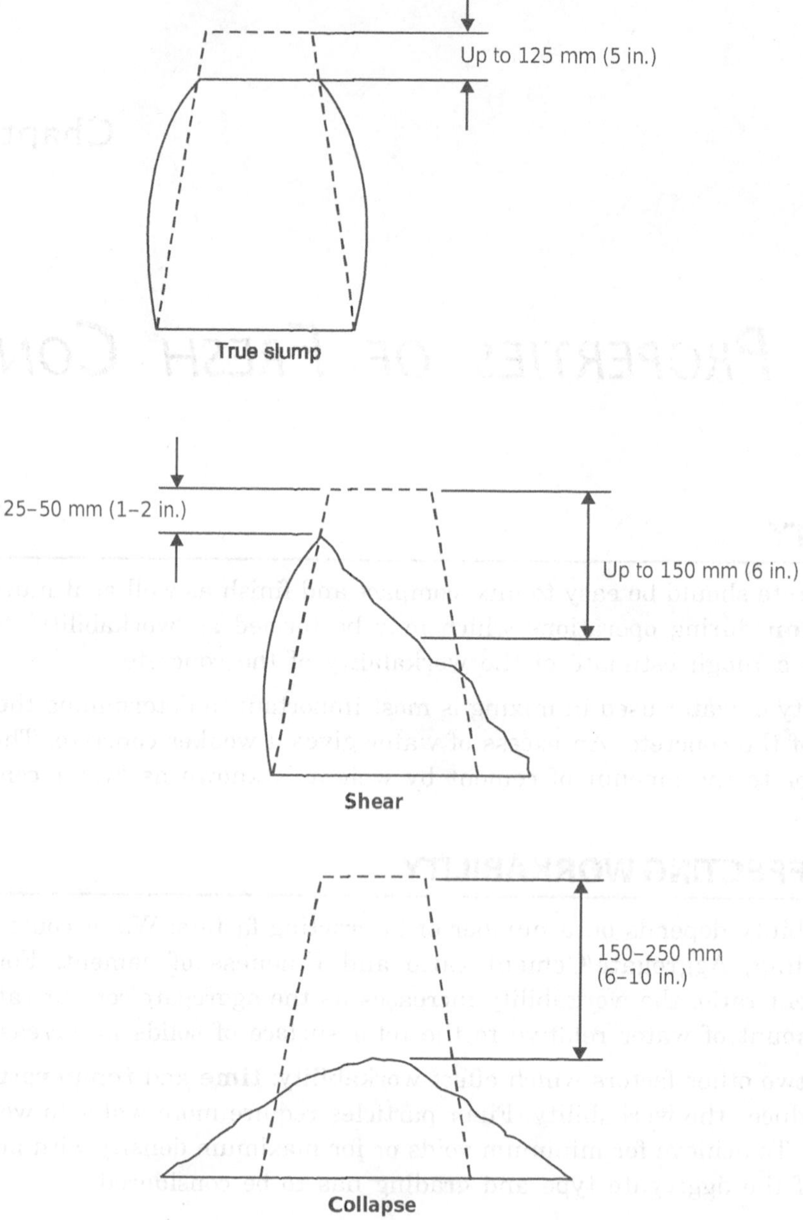

Fig. 4.1. Slump: True, Shear, and Collapse

Immediately after filling the cone is slowly lifted and the unsupported concrete will now slump in mm. The decrease in the height of the slumped concrete is called slump. Suggested ranges of workability of concrete is given as per IS Code of Practice.

Placing conditions (1)	Degree of workability (2)	Slump (mm) (3)
Blinding concrete, Shallow sections, Pavements, etc.	Very low	0.75 to 0.80
Mass concrete, Reinforced sections in slabs, Beams, Walls, Columns, Floors, etc.	Low	25 to 75
Heavily reinforced sections in slabs, Beams, Walls, Columns, etc.	Medium	50 to 100
Trench fill, In situ piling, etc.	High	100 to 150

COMPACTING FACTOR TEST

As shown in Fig. 4.2, the degree of compaction is called the **compacting factor**. It is measured by the density ratio *i.e.,* the ratio of the density actually achieved in the test to the density of the same concrete fully compacted. The apparatus consists of two hoppers, each in the shape of a frustum of a cone about 1.2 m high and one cylinder, the three being above one another. The hoppers have hinged doors at the bottom. All inside surfaces are polished to reduce friction.

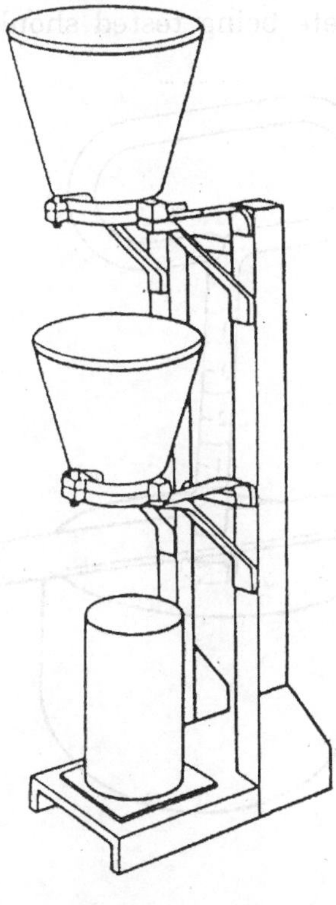

Fig. 4.2. Compacting Factor Apparatus

The upper hopper is filled with concrete. The bottom door of the hopper is then released and the concrete falls into the lower hopper. This hopper is smaller than the upper one. The bottom door of the lower hopper is released and the concrete falls into the cylinder. Excess concrete is cut by slid across the top of the mould and the net mass of concrete in the known volume of the cylinder is determined.

The density of the concrete in the cylinder is calculated. The ratio of this density to the density of the fully compacted concrete is defined as the compacting factor. Compacting factor of concretes with 20 mm or 40 mm maximum size of aggregate is shown below.

Degree of workability	Slump (mm)	Compacting factor
Very low	0 to 25	0.78
Low	25 to 50	0.85
Medium	25 to 100	0.92
High	100 to 175	0.95

KELLY BALL TEST

As shown in Fig. 4.3, this is a simple field test of the determination of the depth to which a 152 mm diameter metal hemisphere weighing 14 kg will sink under its own weight into fresh concrete. The use of this test is similar to that of the slump test. It is for routine checking of consistence for control purposes. The Kelly ball test as an alternative to the slump test. The depth of the concrete being tested should not be less than 200 mm.

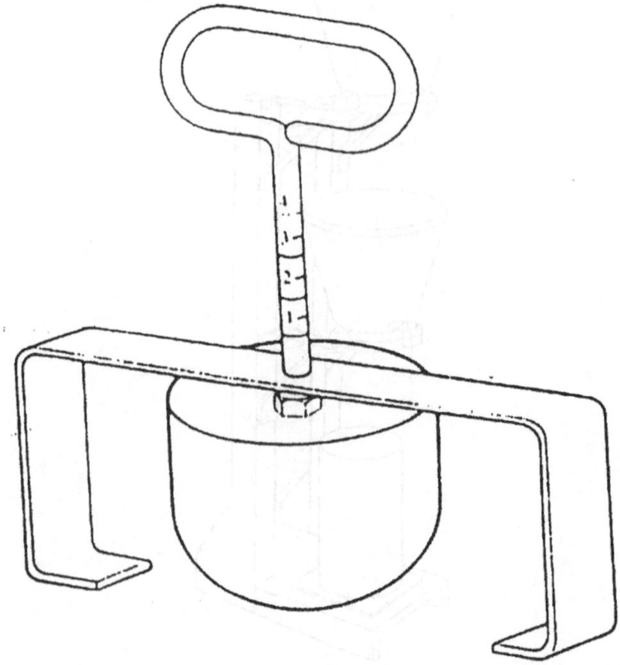

Fig. 4.3. Kelly Ball

FLOW TABLE TEST

As shown in Fig. 4.4, this test has recently become more wide – spread in its use. The apparatus consists a wooden board covered by a steel plate with a total mass of 16 kg. This board is hinged along one side to a base board (700 mm square). The upper board can be lifted the free edge 40 mm. The total top is moistened and a frustum of cone of concrete is placed. After an interval of 30 second the mould is slowly removed. The concrete spreads and the maximum spread parallel to the two edges of the table is measured. The average of these two values, given to mm represents the flow. A value of 400 indicates a medium workability and 500 a high workability.

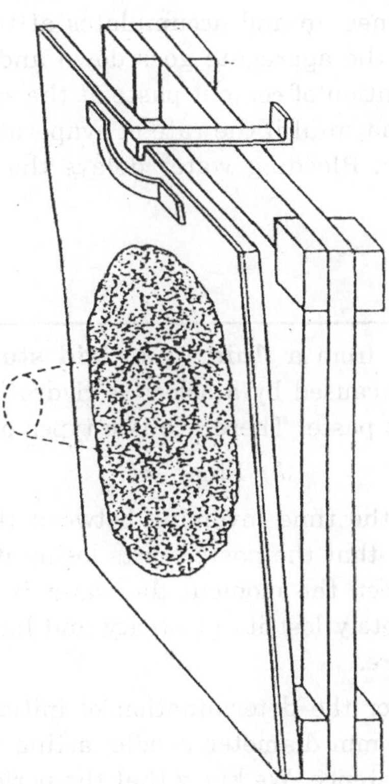

Fig. 4.4. Flow Table Test

SEGREGATION

In considering the workability of concrete, it is pointed out that concrete should not segregate. Segregation can be defined as the separation of the constituent materials of concrete. A good concrete is one in which all the ingredients are properly distributed to make a homogeneous mixture. If a sample of concrete exhibits a tendency for separation of say, coarse aggregate from the rest of the ingredients, then that sample is said to be showing the tendency for segregation. Such concrete is to be weak.

Segregation may be of three types—*(1)* the coarse aggregate separating out from the rest of the matrix, *(2)* the paste separating away from coarse aggregate and *(3)* water separating out from the rest of the material.

Segregation is difficult to measure quantitatively but it can be easily observed at the time of concreting operation. **Slump test** or **Flow Table test** gives a fair idea of the quality of concrete with respect to segregation.

BLEEDING

It is known as *water gain* in a form of segregation in which some of the water in the maximum tends to rise to the surface of the concrete. Bleeding is observed in a highly wet mix. Due to bleeding, water comes up and accumulates at the surface. When the surface is worked up with the **trowel**, the aggregate goes down and the cement and water come up to the top surface. This formation of cement paste at the surface is known as **Laitance.** The bleeding is not completely harmful if the rate of evaporation of water from the surface is equal to the rate of bleeding. Bleeding water delays the texturing and application of curing compounds.

SETTING TIME

Setting refers to a change from a **fluid** to a **rigid state**. It is the stiffening of the cement paste. Setting is mainly caused by a selective hydration of C_3A and C_3S and the temperature rises in the cement paste. There are two types of setting of the cement paste.

(*a*) Initial set and (*b*) Final set.

The initial setting time is the time involving between the moment that the water is added to the cement to the time that the paste starts losing its plasticity. The final setting time is the time involving between the moment the water is added to the cement and the time when the paste has completely lost its plasticity and has attained sufficient firmness to resist certain definite pressure.

(*a*) Initial setting time: For the determination of initial set, the Vicat apparatus is used, this time with a 1 mm diameter needle, acting under a prescribed weight on a paste of standard consistence. We know that the period involving between the time when water is added to the cement and the time at which the needle penetrates the test block to a depth equal to 33–35 mm from the top is taken as initial setting time.

(*b*) Final setting time: This time the Vicat apparatus is with a circular attachment with a centre needle. The cement shall be considered as finally set when lowering the attachment gently cover the surface of the test block, the centre needle makes an impression, while the circular cutting edge of the attachment fails to do so. The paste of the cement has attained such hardness that the centre needle does not go through the paste more than 0.5 mm. British Standards prescribe the initial setting time as a minimum time of 45 minutes and the final setting time as a maximum of 10 hours for Portland cements.

MIXING AND VIBRATION OF CONCRETE

The mixing operation consists of **rotation** or **stirring** of the ingredients to coat the surface of all the aggregate particles with cement paste and to blend all the ingredients of concrete into a uniform mass. The correct quantities of cement, aggregates, water and if admixture require, are batched and mixed in a concrete mixer so that the mixer is suitable for mixing for the workability of the concrete.

The process of compacting concrete by vibration consists of the elimination of **entrapped air** and **forcing** the particles into a closer configuration. Both compaction by hand and compaction by vibration can produce good quality concrete, with the right mix and workmanship. The poker vibrator or immersion vibrator is used for the process of compacting concrete.

MIXERS AND VIBRATORS

The usual type of mixer is a batch mixer. It means that one batch of concrete is mixed and discharged before any more materials are put into the mixer. There are four types of batch mixers— *(1)* **A tilting drum mixer** is one whose drum is tilted for discharging, *(2)* **A non-tilting drum mixer** is one in which the axis of the mixer is always horizontal, *(3)* **A pan-type mixer** is a forced action mixer and *(4)* **A dual drum mixer** is sometimes used in highway construction.

Internal vibrators is the most common one. It consists of a poker housing a shaft driven through a flexible drive from a motor. The poker is immersed in concrete and thus applies forces to it. The poker can be easily moved from place to place so that the concrete is vibrated in it.

External vibrators are rigidly clamped to the formwork which rests on an elastic support so that both the form and the concrete are vibrated.

METHODS OF CURING

Once the concrete has set, wet curing can be provided by keeping the concrete in contact with water. This may be achieved by spraying or flooding (ponding) or by covering the concrete with wet sand, earth, sawdust or straw. Periodically wetted Hessian or cotton mats can be used. A continuous supply of water is more efficient than periodically wetted one. The strength development of concrete depends on water curing. It is advisable to commence curing as soon as possible because in practice early drying may lead to **shrinkage** and **cracking** of the concrete. Minimum period of curing and protection for surface temperature of concrete 21 days to 28 days.

MATURITY

Since the strength development of concrete depends on both **time** of curing period and **temperature** on the gain of strength. It can be said that strength is a function of summation of product of time and temperature. This summation is called **maturity** of concrete.

$$\text{Maturity} = \Sigma\ (\text{time} \times \text{temperature})$$

It is found that the hydration of concrete continues to take place up to about –11°C (12°F). Therefore, –11°C is taken as a datum line for computing maturity. Maturity is measured in degree centigrade hours (°C hrs) or degree centigrade days (°C days).

SOLVED QUESTION

Q. 1. A sample of concrete cured at 18°C for 28 days is taken as fully matured concrete.

Sol. Its **maturity** is equal to 28 × 24 × [18 – (11)] = 19488°C h.

SUMMARY

- Fresh concrete should be easy to mix, compact and finish as well as it must remain free from segregation during operations which may be termed as 'workability'.
- Slump test, is the most commonly used method of measuring consistency of concrete.
- The workability depends on a number of interacting factors: Water content, Aggregate type and Grading, Aggregate/Cement ratio and Fineness of cement.
- The mixing operation consists of rotation or stirring of the ingredients to coat the surface of all the aggregate particles with cement paste and to blend all the ingredients of concrete into a uniform mass.

REVIEW QUESTIONS

1. What is slump test and compacting factor test?
2. What is Kelly ball test and Flow table test?
3. What is segregation and bleeding?
4. How many types setting of the cement test? Explain.
5. How many types of batch mixers?

Chapter 5

STRENGTH OF CONCRETE

WATER/CEMENT RATIO

The strength of concrete depends on the effective water/cement ratio. It is calculated on the basis of the mix water less the water absorbed by the aggregate. Strength of concrete primarily depends upon the strength of cement paste. The strength of **paste** increases with **cement content** and decreases with **air** and **water content.**

Thus, water/cement ratio law states that the strength of cement is only dependent upon water/cement ratio provided the mix is **workable**.

GEL/SPACE RATIO

The **gel/space ratio** is defined as the ratio of the volume of the hydrated cement (paste) gel to the sum of the volumes of the hydrated cement and of the capillary pores.

The **strength of concrete** = $240x^3$ where x = the gel/space ratio and 240 = the strength of the gel (paste) in MPa or N/mm^2 .

Calculation of Gel/Space Ratio for Complete Hydration

$$\text{Gel/Space ratio } = X = \frac{\text{Volume of Gel}}{\text{Space Available}} = \frac{0.657C}{0.319C + W}$$

where,

C = Weight of Cement in g.

V_c = 0.319 ml/gm = Specific Volume of Cement.

W_o = Volume of Mixing Water in ml.

Assuming that 1 ml of cement on hydration will produce 2.06 ml of gel.

$$\text{Volume of gel} = C \times 0.319 \times 2.06$$

$$\text{Space available} = C \times 0.319 + W_o$$

Calculation of Gel/Space Ratio for Partial Hydration

$$\text{Gel/Space ratio} = X = \frac{\text{Volume of Gel}}{\text{Space Available}}$$

$$= \frac{2.06 \times 0.319 \times \alpha \times C}{0.319C \times \alpha + W_o}$$

where, α = Fraction of Cement that has hydrated

$$\text{Volume of gel} = C \times \alpha \times 0.319 \times 2.06$$

$$\text{Total space available} = C\, V_c\, \alpha + W_o$$

Solved Problem

Q. 1. Calculate the gel/space ratio and the theoretical strength of a sample of concrete made with 500 gm. of cement with 0.5 water/cement ratio, on full hydration and at 60% hydration.

Sol.

(i) **Gel/space ratio on full hydration:**

$$X = \frac{\text{Volume of Gel}}{\text{Space Available}} = \frac{0.657C}{0.319C + W}$$

$$= \frac{0.657 \times 500}{0.319 \times 500 + 0.5 \times 500} = \frac{328}{160 + 250}$$

$$= \frac{328}{410} = 0.8$$

Theoretical **strength of Concrete**

$$= 240 \times (0.8)^3 = 123 \text{ MPa or N/mm}^2.$$

(ii) **Gel/space ratio for 60% hydration:**

$$\text{Gel/Space ratio} = X = \frac{\text{Volume of Gel}}{\text{Space Available}} = \frac{2.06 \times 0.319 \times \alpha \times C}{0.319C \times \alpha + W_0}$$

$$= \frac{0.657 \times 500 \times 0.6}{0.319 \times 500 \times 0.6 + 250} = \frac{197.1}{345.7} = 0.57$$

Theoretical **strength of Concrete** at 60% hydration = $240 \times (0.57)^3 = 44.4$ MPa.

STRENGTH IN TENSION

It is useful to determine the tensile strength of concrete. This test is covered under the **American Society for Testing Machine Standard (ASTM) C 307.** The tensile strength for good Portland Cement should be not less than 2 N/mm^2 after 3 days and not less than 2.5 N/mm^2 after 7 days.

STRENGTH IN COMPRESSION

Compression test is the most common test conducted on hardened concrete. Because it is an easy test to perform to its compressive strength.

It is determined using **cube specimen** of the size **15 × 15 × 15 cm**. It is cast in a mould, preferably with a clamped base, made from steel, **cast iron**, brass, etc., which satisfy the **physical requirements** of water tightness, absorptive and elongation.

A thin layer of **mineral oil** has to be applied to the **inside** surfaces of the moulds in order to **prevent** bond between the concrete and the mould. Concrete is then placed in the mould in **layers**. Compaction of high–slump layer being compacted by 25 strokes of a 16 mm diameter steel rod.

The top surface of a cube, finished by a **trowel**, is not plane and smooth enough for testing. In addition to being plane, the end surfaces of the test cube should be normal to its axis. The testing machine is equipped with a seating which can align freely. The **curing** conditions for the standard test cubes are specified by the BIS.

When cast in the laboratory, the molded specimen are stored in **water** so that moisture loss is prevented for certain period. The most common **age** at test is **28 days**, but additional tests can be made at 3 and 7 days.

The cube is then placed with the forces in contact of the testing machine. The load is applied at a constant rate of stress. Due to compression load, the **cube** undergoes **lateral expansion** owing to the **Poisson's ratio effect**. While the vertical load is imposed compression, there is a **lateral shearing stress** as well as **compression** which increases the strength of cubes.

The crushing strength is the required compressive strength. The maximum **recorded load** divided by the **area** of cross section of the **specimen** gives the **compressive strength.**

EFFECT OF AGE ON STRENGTH

The concrete develops **strength** with **hydration**. With an increase in **age,** the degree of hydration increases so that strength increases. The rate of gain of **strength** is **faster** to **start with** and the rate gets **reduced** with **age**. It is assumed that the **28 days strength** as the **full strength** of concrete. The concrete develops strength **beyond** 28 days also.

The Codes have not been permitting to the strength beyond 28 days for design purposes. The increase in strength beyond 28 days is used as the **factor of safety**.

RELATION BETWEEN COMPRESSIVE AND TENSILE STRENGTH

The theoretical compressive strength is stated to be **eight times** larger than the tensile strength. The tensile strength increases with **age** at a lower rate than the compressive strength. However, there are several **other factors** which affect the relation between the two strengths. The main factors are the method of testing the concrete in tension, the size of the specimen, the shape and surface texture of coarse aggregate and the moisture condition of the concrete.

In reinforced concrete construction the strength of the concrete in compression is only **taken into consideration**. The tensile strength of concrete is generally **not taken into consideration.**

There are number of **empirical relationships** connecting tensile strength and compressive strength of concrete. One of the **common relationships** is shown below:

(*i*) **Tensile Strength = K (Compressive Strength)^n** where, Value of **K** varies from **6.2 for gravels to 10.4** for crushed rock (average value is **8.3**) and value of ***n*** may vary from **0.5 to 0.75.**

(*ii*) The Indian Standard **IS : 456 of 2000** gives the following relationship between the compressive strength and flexural strength.

$$\text{Flexural Strength} = 0.7\sqrt{f_{ck}}$$

where f_{ck} is the characteristic compressive strength of concrete in N/mm^2.

FATIGUE STRENGTH

Two types of **failure in fatigue** can take place in concrete.

(*i*) In the first failure occurs under a **sustained load** (or a **slowly** increasing load) near, but below the strength under an increasing load as in a standard test. This is known as **static fatigue** or **creep rupture.**

(*ii*) The second type of failure occurs under **cyclic** or **repeated loading** and it is known simply as **fatigue.**

In both cases, a **time-dependent failure** occurs only at stresses which are greater than a certain **threshold value** but smaller than the **short-term static strength.**

In the standard method, the compressive strength is determined in a test of short duration *viz.* 2 to 4 min. The duration of the test is important because strength is dependent upon the rate of loading.

Static fatigue also occurs in tension at stresses greater than 70 to 80% of the short-term strength but the tensile strain capacity is much lower than in compression.

STRESS-STRAIN RELATION AND MODULES OF ELASTICITY

The stress-strain behaviour of Concrete depends on a number of variables such as the properties of the materials with which concrete is made and the loading parameters such as the rate of loading and the age at testing. The Cement paste as well as the aggregates exhibit linear stress-strain properties. While the concrete behaves as non–linear of the stress-strain curve.

The modulus of elasticity is determined by subjecting a cube specimen to uniaxial compression and measuring the deformations by means of dial gauges fixed between certain gauge length. Dial gauge reading divided by gauge length will give the strain and load applied divided by area of cross-section will give the stress. Thus, the stress-strain relationship is established.

POISSON'S RATIO

Poisson's ratio is defined as the ratio of lateral strain to axial strain (or longitudinal strain) under uniaxial loading within the elastic range. For Concrete, the value of Poisson's ratio generally lies between 0.15 and 0.20.

SHRINKAGE AND CREEP

Shrinkage can be defined as the decrease in the **volume** of concrete with **time**. This decrease is due to changes in **moisture content**. These changes occur without stress.

When a viscous elastic material is subjected to **stress** for a particular **duration**, its **strain** changes over **time**. This time-dependent increase/ decrease in the strain of hardened concrete subjected to a sustained stress is termed **Creep.**

Creep is closely related to shrinkage. Both Creep and shrinkage depend on the behaviour of hydrated cement paste. Concrete that is resistant to shrinkage is also resistant to creep.

COMPRESSION TEST ON CUBES AND CYLINDERS

The Compression test is used to determine the **hardness of Cubical and Cylindrical specimens** of Concrete. The strength of a concrete specimen depends upon **cement, aggregate, bond, W/C ratio, curing, temperature, age and size of specimen.** Mix design is the **major factor** controlling the strength of concrete Cubes of size 15 cm × 15 cm × 15 cm (as per **IS : 10086–1982**) should be cast. The specimen should be given sufficient **time** for **hardening** (approx. 24 h) and then it should be cured for 7 days and 28 days. After 7 days or 28 days, it should be loaded in the **compression testing machine** and tested for maximum load. Compressive strength should be calculated by dividing maximum **load** by the cross-sectional **area**.

The Compressive strength is used for establishing the type of concrete by routine testing by making cubes at the construction site and testing them at the end of 28 days. By conducting tests on a **cube**, only **strength parameters** can be obtained.

The compressive strength of concrete is assessed by testing (150 mm × 300 mm) **cylinders**. Testing cylinders enables one to obtain the **stress-strain properties** of concrete under uniaxial compression.

NON-DESTRUCTIVE TESTS

It is as follows.

Rebound Hammer Test

This test can be used to evaluate the compressive strength of the affected portion and the result can be correlated with the designed strength of the elements of the structure.

Ultrasonic Pulse-Velocity Test

This test can be used to **investigate** the extent and depth of deep-rooted cracks. In this test, an **ultrasonic pulse** is sent with the help of a **transmitter** and received on the opposite face with the help of a **receiver**. The measurement of the **velocity of sound** can indicate the **detour** taken by the **sound wave** due to the presence of cracks in the structure and the location/depth of the crack can be monitored on a **cathode ray oscilloscope.** The effect of **internal cracking** is obtained on the display in an **oscilloscope screen**.

Potential Measurement

Under favourable conditions, it is possible to use **electrical measurements** from the surface of concrete to get an indication of the condition of steel inside. Electrical connections have to be made with a **rebar** by making a small hole in the concrete. In porous concrete such a hole may not be required as wetting by the electrical connection with concrete is made by means of an **electrolyte** which wets both concrete and the conductor.

SUMMARY

- Poisson's ratio is defined as the ratio of lateral strain to axial strain (or longitudinal strain) under uniaxial loading within the elastic range.
- The gel/space ratio is defined as the ratio of the volume of the hydrated cement (paste) gel to the sum of the volumes of the hydrated cement and of the capillary pores.
- Compression test is the most common test conducted on hardened concrete. Because it is an easy test to perform to its compressive strength.
- The theoretical compressive strength is stated to be eight times larger than the tensile strength.

REVIEW QUESTIONS

1. What is difference between water/cement ratio and gel/space ratio?
2. What is strength in compression?
3. What is the relation between compressive and tensile strength?
4. What is shrinkage and creep?
5. What is non-destructive tests?

Chapter 6

ADMIXTURES

INTRODUCTION

Admixture is defined as a material, other than cement, water and aggregates. It is used as an ingredient of concrete. Admixtures are added to modify the important properties of concrete in the fresh or hardened state of concrete.

DIFFERENT TYPES, EFFECTS AND USES

Main classification of admixtures are given below:

Type of admixture	Effect produced on concrete
Accelerators	Accelerating and Water-reducing
Retarders (Water-reducing)	Retarding and Water-reducing
Plasticizers (Air-entraining)	Air–entraining, Water-reducing and increasing workability.
Superplasticizers	Flowing Concrete increasing workability inducing flowability.
Waterproofers	Damp proofing.

RETARDERS

Retarders increase the **setting time** of the concrete mix and reduce the **water/cement ratio**. Usually up to **10% water reduction** can be achieved. A wide range of **water-reducing and set-retarding admixtures** are used in ready mixed concrete. They have a detergent-like property.

Retarders are useful when concreting in **hot weather**, when the normal setting time is shortened by the higher temperature. The chemicals are derived from lignosulphonic acids and their salts, sulphonated melamine or naphthalene form aldehyde.

SUPER PLASTICIZERS

These are more recent and more effective type of **water-reducing admixtures**. It produces extreme **workability** and thus flowing concrete. They achieve reduction in the water content without loss of workability. Their use generally leads to an overall reduction in the cost.

Super plasticizer molecules and cement grains are oppositely charged and hence repel each other. This increases the mobility and hence makes the concrete flow. Super plasticizer enable savings in cement for a given strength and are ideal for pumping concrete, casting heavily reinforced concrete members and the pre-cast elements of concrete.

MIX DESIGN BY IS CODE METHOD

Considerable need has been felt for formulating standard recommendations for guidelines on proportioning of **concrete mixes**. The need has been further emphasized by the importance given to design mix concrete according to **IS : 10262–1982.** It is the **recommended procedure for designing concrete mixes** for general types of construction using the concreting materials normally available. The design is carried out for a desired **Compressive strength and workability** of concrete using graded aggregates.

The following procedures are shown for a **Concrete Mix Design for M 25 Grade Concrete:**

GHERZI EASTERN LIMITED

16, Mahanirban Road, Kolkata–700 029

July 20, 2007

Calculation sheet for trial Mix M 25 Grade Concrete

The Final Mix Proportion: Brand of Cement—Ultra Tech Cement (PPC), Grade 53

Cement (kg)	Water (Ltrs.)	Sand (F.A.) kg.	20 mm–70%	12.5 mm–30%
400	160	621.76	870.22	372.95
50	20	77.72	108.78	46.62
1	0.40	1.55	2.18	0.93
18	7.20	27.98	39.16	16.78

Actual quantities are required for the mix per 18 kg of cement, date of made 27 W 07 M 2007 Year, after water absorption:

Concrete batch no. 1

Cement = 18 kg
Water = 8.00 ltrs
Sand = 27.62 kg
20 mm stone chips = 38.85 kg
12.50 mm stone chips = 16.65 kg
Water Cement ratio = 0.40
Admixture = No Admixture is added.

Target strength of concrete at 28 days = **31.60 N/mm^2**

No. of cube taken for tests : 6 Nos. only

Weather: Sunny/Cloudy/Rainy

Mixing apparatus: Diesel Mixer Machine

Slump observed: 90 mm (Vibrator is not used)

7 Days Test Result, casted on 20.07.2007, tested on 27.07.2007.

Cube mark	Crashing load	Ave. compressive strength
M(25)/1	480 kN	
M(25)/2	410 kN	18.37 N/mm^2
M(25)/3	350 kN	

28 Days Test Result, casted on 20.07.2007, tested on 17.08.2007.

Cube mark	Crashing load	Ave. compressive strength
M(25)/4	840 kN	
M(25)/5	650 kN	31.85 N/mm^2
M(25)/6	660 kN	

Witnessed by:

For Consultant	For Agency	For Client
M/s G E Ltd.	M/s Sen Bro.	M/s L&T Ltd.

GHERZI EASTERN LIMITED

16, Mahanirban Road, Kolkata - 700 029

Kolkata Regional Office.

Report on Concrete Mix Design for M 25 Grade Concrete

July 20, 2007

It is carried out for **Construction of proposed expansion of Service Station and Office Building** of M/s **Larsen and Tourbo Ltd.** at Degaul Avenue, Khoirasol, Durgapur - 713 212.

Name of Consultant

M/s **GHERZI Eastern Ltd.**
Kolkata Regional Office
16, Mahanirban Road
Kolkata - 700 029.

Name of Agency

M/s **Sen Brothers**
Angadpur
Durgapur - 713 215.

Design Data

(1) Degree of workability : 0.8 compacting factor
(2) Degree of quality control : Very good
(3) Type of expose : Moderate

Test Data for Materials

The following materials are used in **trial mixes**

(1) **Type of cement**:
(*a*) Brand - **Ultra Tech Cement 53**
Portland Pozzolana Cement (**PPC**)
Fly ash base, IS : 1489 (Part 1) - 1991,
Made on 27 W 7 M 2007 Y.
(*b*) Specific gravity : 2.87

(2) **Aggregate**: (*a*) Specific gravity of coarse aggregate: 2.81

(*b*) Specific gravity of find aggregate: 2.61

(3) **Water absorption**: (*a*) Coarse aggregate: 0.8%

(*b*) Fine aggregate: 1.3%

(4) **Free moisture**: (*a*) Coarse aggregate: Nil

(*b*) Fine aggregate: Nil

Sieve analysis of coarse aggregate taking 70%–20 mm & 30%–12.5 mm:

IS Sieve	% Passing	% Passing	Total % passing	Limit as per IS: 383, Table–2
	20 mm – 70%	12.5 mm – 30%		
40 mm	70.00	30.00	100	100
20 mm	69.50	30.00	99.50	95–100
10 mm	2.60	26.10	28.70	25–55
4.75 mm	0.32	5.58	5.90	0–10

Gradation of Sand (Sieve Analysis of Fine Aggregate):

IS Sieve	Weight (in gms) retained	%Weight retained	Cumulative % retained	% passing	Limit as per IS: 383, Table – 4
10 mm	0	0	0	100	100
4.75 mm	76	7.60	7.60	92.40	90–100
2.36 mm	39	3.90	11.50	88.50	85–100
1.18 mm	124	12.40	23.10	76.10	75–100
600 Mic.	117	11.70	34.80	65.20	60–79
300 Mic.	505	50.50	85.30	14.70	12–40
150 Mic	95	9.55	94.85	5.15	0–10

Total = 257.15 = Sum of cum % retained

Sand confirm to Zone III as per IS – 383, Fineness modulus of F.A.

= (Sum of Cum % retained/100) = 2.57

Characteristic compressive strength is required in the field at 28 days = 25 N/mm^2.

Target strength of concrete at 28 days = 25 + 1.65 × 4 = **31.60 N/mm^2**

Maximum size of aggregate = 20 mm.

It is tried with water cement ratio = 0.40

Considering for one cum of concrete

Cement content = 400 kg per cum

Water content = 0.40 × 400 = 160 ltrs.

Fine aggregate (Sand) is taken as 35% of the total aggregate to get cohesiveness. For the specified maximum size of aggregate of 20 mm, the amount of entrapped air in the wet concrete is 2% as per IS : 10262 – 1982.

Determination of coarse and fine aggregate content:

Applying equation from 3.5.1 of IS : 10262

Fine aggregate (Sand):

$$0.98 = \{160 + 400/2.87 + fa/(2.61 \times 0.35)\} \times 1/1000$$

$$\text{Or } fa = 680.63 \times 2.61 \times 0.35$$

$$= 621.76 \text{ kg}$$

Coarse aggregate:

$$0.98 = \{160 + 400/2.87 + Ca/(2.81 \times 0.65)\} \times 1/1000$$

$$\text{Or } Ca = 680.63 \times 2.81 \times 0.65$$

$$= 1243.17 \text{ kg}$$

20 mm of 70% = 870.22 kg

12.5 mm of 30% = 372.95 kg

The Final Mix Proportion then becomes:

Cement (kg)	Water (Ltrs.)	Sand (F.A.) kg.	20 mm – 70% kg.	12.5 mm – 30% kg.
400	160	621.76	870.22	372.95
50	20	77.72	108.78	46.62
1	0.40	1.55	2.18	0.93
18	7.20	27.98	39.16	16.78

Actual quantities are required for the mix per 18 kg of cement after water absorption:

Quantity of sand for dry condition

$$= 27.98 - (27.98 \times 1.3\%)$$

$$= 27.98 - 0.36 = 27.62 \text{ kg}$$

Quantity of 20 mm coarse aggregate for dry condition

$$= 39.16 - (39.16 \times 0.8\%)$$

$$= 39.16 - 0.31 = 38.85 \text{ kg}$$

Quantity of 12.50 mm coarse aggregate for dry condition

$$= 16.78 - (16.78 \times 0.8\%)$$

$$= 16.78 - 0.13 = 16.65 \text{ kg}$$

Quantity of water to be added

$$= 7.20 + 0.36 + 0.31 + 0.13 = 8.00 \text{ ltrs.}$$

Therefore, the actual quantities of different constituents required for the mix are:

Cement = 18 kg
Water = 8.00 ltrs
Sand = 27.62 kg
20 mm stone chips = 38.85 kg
12.50 mm stone chips = 16.65 kg

P. K. Nag
Sr. Project Engineer
G. E. Ltd.

LIGHT-WEIGHT CONCRETE

The conventional concrete is the **high self weight** of concrete. Density of the normal concrete varies from **2200 to 2600 kg/m³**. This heavy self weight will make **uneconomical** structural material. The light-weight concrete as we call is a concrete whose density varies from **300 to 1850 kg/m³**.

There are many **advantages** of having low density. It helps in **reduction** of dead load, increases the progress of building. If floors and walls are made up of light-weight concrete it will result in considerable **economy**. The light-weight concrete is achieved in practice by **three** different ways. These are:

(*i*) By replacing the usual mineral **aggregate** by using **cellular porous or light-weight aggregate.**

(*ii*) By introducing **gas, air bubbles or large voids** within the concrete or mortar mass. This is known as **aerated** concrete.

(*iii*) By omitting the fine (sand) aggregate from the mix so that a large number of voids is present. This is called **'no – fines'** concrete.

POLYMER CONCRETE

Just as cement is used as a binder in cement concrete, **monomer** or **resin** is added to bind aggregates of coarse, fine and other particle. A monomer is an organic molecule which is capable of combining chemically with different molecules, known as **Polymer.** The commonly used binders for Polymer concrete are **styrene, methylmethacrylate, polyesters** and **epoxies.**

In the premix method polymer and aggregate are mixed in conventional mixers and the mix is transferred to moulds. The mix is vibrated for compaction. Concretes with **polymers** are added during mixing to modify the properties of the hardened concrete. A wide range of **monomers** are being used in concrete.

The prices of various polymers vary considerably but as a rough guide it can be assumed that polymers cost approximately 20 times that of ordinary cement.

FIBRE-REINFORCED CONCRETE

Fibre-reinforced concretes have improved **tensile strength** and **toughness** compared to conventional concrete.

Plain concrete possesses a very low tensile strength and little resistance to **cracking**. Internal **micro cracks** are found present in the concrete.

Fibre-reinforced concrete has proved to be a good **alternative** to conventional concrete. Fibre-reinforced concrete is defined as concrete made with **hydraulic cement** containing **fine or fine and coarse aggregate** and **fibres**. The **fibres** can be made from **natural material** (e.g., **asbestos, sisal, cellulose**) or from a manufactured product such as **glass, steel, carbon and polymer** (e.g., **polypropylens, Kevlar**). The purposes of reinforcing the cement-based matrix with fibres are to increase the **tensile strength** by delaying the **growth of cracks.**

The quantity of fibres **used** is small which is **1 to 5%** by volume. Fibre is a small piece of reinforcing material possessing certain **characteristic properties**. They can be circular or flat. **Steel fibre** is one of the most commonly used fibre.

SOLVED QUESTIONS

Q. 1. What do you understand by the grade of concrete mixes?

Sol. The grade of concrete mixes are $\mathbf{M_{10}}$, $\mathbf{M_{15}}$, $\mathbf{M_{20}}$, $\mathbf{M_{25}}$, $\mathbf{M_{30}}$, $\mathbf{M_{35}}$, $\mathbf{M_{40}}$, etc. **'M'** refers to the **mix** and the **number** denotes the **ultimate strength** of concrete mix in N/m^2 at the end of 28 days.

Q. 2. What are the **proportions** of the following grade of concrete?

$\mathbf{M_{10}}$, $\mathbf{M_{15}}$, $\mathbf{M_{20}}$ and $\mathbf{M_{25}}$.

Sol. $\mathbf{M_{10} \rightarrow 1 : 3 : 6}$, $\mathbf{M_{15} \rightarrow 1 : 2 : 4}$, $\mathbf{M_{20} \rightarrow 1 : 1\frac{1}{2} : 3}$, and $\mathbf{M_{25} \rightarrow 1 : 1 : 2}$.

Q. 3. What do you understand by **Ingredients of concrete**?

Sol. The **ingredients** of concrete are the following:

(*i*) Cement → Weight **50 kg** per bag → Volume **0.0347 cum.**

(*ii*) Fine aggregate (Sand) → Less than **4.75 mm Sieve size.**

(*iii*) Coarse aggregate (broken stones or gravel) → More than **4.75 mm Sieve size**.

(*iv*) Water

(*v*) Recently, a large number of additives (**Admixture**) known as **concrete additives** are also added as ingredients of concrete to make concrete of qualities required for various constructions. Some of the **additives** used are the following:

(*a*) Plasticizers
(*b*) Superplasticizers
(*c*) Accelerators
(*d*) Retarders
(*e*) Fibres
(*f*) Pozzolanic material

SUMMARY

- Retarders increase the setting time of the concrete mix and reduce the water/cement ratio.
- Considerable need has been felt for formulating standard recommendations for guidelines on proportioning of concrete mixes.
- The conventional concrete is the high self weight of concrete. Density of the normal concrete varies from 2200 to 2600 kg/m^3.
- Just as cement is used as a binder in cement concrete, monomer or resin is added to bind aggregates of coarse, fine and other particle.
- Fibre-reinforced concretes have improved tensile strength and toughness compared to conventional concrete.

REVIEW QUESTIONS

1. What is retarders and super plasticizers?
2. What is mix design by IS code method? Explain.
3. What is light-weight concrete and polymer concrete?
4. What is Fibre-reinforced concrete?

MULTIPLE CHOICE TYPE QUESTIONS

1. The main raw material for the production of Cement is called
   (*a*) Clinker (*b*) Slurry
   (*c*) Hydration (*d*) Hard mass
2. A huge inclined rotating furnace in which the raw materials are used to derive the chemical components of Cement with temperature is known as
   (*a*) Cement furnace (*b*) Cement making unit
   (*c*) Cement Kiln (*d*) Cement process vessel
3. Which material is added to regulate the setting time of Portland Cement during manufacturing process
   (*a*) Limestone (*b*) Stale Clay
   (*c*) Iron Ore (*d*) Gypsum
4. The impure C_3 S is known as in the chemical components of Cement
   (*a*) Residue Calcium silicate (*b*) Belite
   (*c*) Alite (*d*) None of these
5. The impure C_2 S is known as in the chemical components of Cement
   (*a*) Residue Calcium Silicate (*b*) Belite
   (*c*) Alite (*d*) None of these

Some Important Test on Cement/Concrete

TESTS ON CEMENT

EXPERIMENT NO. 01

• **Determination of Specific Gravity of Cement (IS : 2720–Part 3)**

Date: ________

Object

To determine the specific gravity of Cement using Specific Gravity Bottle.

Apparatus

(*a*) Specific Gravity Bottle - 100 ml capacity

(*b*) Balance capable of weighting accurately up to 0.10 gm

Procedure

Weigh a clean and dry Specific Gravity bottle (W_1). Place a sample of cement about 50 gm in the dry Specific Gravity bottle and weight it (W_2). Add Kerosene (polar liquid) to cement about half full. Mix thoroughly with glass rod to remove entrapped air. Continue stirring and add more Kerosene till it is with the graduated mark. Dry the outside and weigh (W_3). Entrapped air may be removed by vacuum pump if available. Empty the bottle, clean it, refills with clean Kerosene with the graduated mark wipe dry the outside and weigh (W_4).

Calculations

$$\text{Specific Gravity of Cement} = \frac{(W_2 - W_1) \times 0.773}{(W_2 - W_1) - (W_3 - W_4)}$$

where, W_1 = Weight of empty bottle

W_2 = Weight of empty bottle + Cement

W_3 = Weight of empty bottle + Cement + Kerosene

W_4 = Weight of empty bottle + Kerosene

G_k = 0.773 = Specific Gravity of Kerosene

Limit

Specific Gravity of Cement = 2.85 to 3.15

EXPERIMENT NO. 02

- **Determination of Fineness of Cement (IS : 4031–Part 3)**

Date: ________

Object

To determine the fineness of Cement using dry sieving.

Apparatus

(*a*) Standard balance with 100 gm weighing capacity.

(*b*) IS : 90 micron sieve confirming to IS : 460–1962 and a brush.

Procedure

(a) Weigh accurately 100 gms of the cement and place it on a standard 90 micron IS sieve.

(*b*) Continuously sieve the sample for 15 minutes.

(*c*) Weigh the residue left after 15 minutes of sieving. This completes the test.

Result

The percentage weight of residue over the total sample is reported.

$$\% \text{ Weight of Residue} = \frac{\text{Wt. of sample retained on the sieve}}{\text{Total Weight of the sample}}$$

Limits

The percentage residue should not exceed 10%. In commercial cement it is suggested that there should be about 25–30% of particles of less than 7 micron in size.

Precautions

Sieving shall be done holding the sieve in both hands with gentle motion. This will involve no danger of spilling the cement, which shall be kept well spread out on the screen. More or less continuous rotation of the sieve shall be carried out throughout sieving. The underside of the sieve shall be lightly brushed with a 25 or 40 mm bristle brush after every five minutes of sieving.

EXPERIMENT NO. 03

- **Determination of Soundness of Cement (IS : 4031–Part 3)**

Date: ________

Object

Determination the Soundness of Cement using Le-Chatelier apparatus.

Apparatus

The apparatus for conducting the test consists of small split cylinder of spring brass or other suitable metal of 0.5 mm thickness, forming a mould 30 mm internal diameter and 30 mm high. On either side of the split mould are attached two indicators with pointed ends, the distance from these ends to the center of the cylinder being 165 mm. The mould shall be kept in good condition with the jaws not more than 0.50 mm apart.

Procedure

Place the mould on a glass sheet and fill it with cement paste formed by gauging cement with 0.78 times the water required to give a paste of standard consistency. The paste shall be gauged in the manner and under the conditions prescribed in determination of consistency of standard cement paste, taking care to keep the edges of the mould gently together while this operation is being performed. Cover the mould with another piece of glass sheet, place a small weight on this covering glass sheet and immediately submerge the whole assembly in water at a temperature of 27° C ± 2° C and keep there for 24 hours.

Measure the distance separating the indicator points. Submerge the moulds again in water at the temperature prescribed above. Bring the water to boiling, with the mould kept submerge in 25 to 30 minutes and keep it boiling for three hours. Remove the mould from the water, allow it to cool and measure the distance between the indicator points. The difference between these two measurements represents the expansion of the cement.

Limits

Expansion should not be more than 10 mm.

Result

Initial Distance of Indicator points (S1)

Final Distance of Indicator points (S2)

Difference between two measurements i.e, Expansion of the cement.

EXPERIMENT NO. 04

- **Determination of Normal Consistency of Cement (IS : 4031–Part 4) 1988**

Date: ________

Introduction

The standard consistency of a cement paste is defined as that consistency which will permit by using the Vicat plunger to penetrate to a depth of 33–35 mm from the top of the mould. The Vicat apparatus is used to find out the percentage of water required to produce a cement paste of standard consistency.

Apparatus

Vicat Apparatus conforming to IS : 5513–1976.

Balance of capacity 1 kg and sensitivity to 01 gram.

Gauging trowel conforming to IS : 10086–1982.

Procedure

Unless otherwise specified this test shall be conducted at a temperature 27°C ± 2°C and the relative humidity of laboratory should be 65 ± 5%.

Prepare a paste of weighed quantity of cement (300 gms) with weighed quantity of potable or distilled water, taking care that the time of gauging is not less than 03 minutes nor more than 05 minutes and the gauging is completed before any sign of setting occurs.

Fill the vicat mould with this paste resting upon a non-porous plate.

Smoothen the surface of the paste, making it level with the top of the mould.

Slightly shake the mould to expel the air.

Lower the plunger gently to touch the surface of the test block and quickly release, allowing it sink into the paste.

Record the depth of penetration.

Prepare trial pastes with varying percentages of water and test as described above until the plunger is 33–35 mm from the top of the vicat mould.

Calculations

$$\text{Standard consistency (\%)} = \frac{\text{Weight of water added}}{\text{Weight of cement}} \times 100$$

Report

Express the amount of water as a percentage by mass of dry cement to the first place of decimal.

Precautions

Care shall be taken to maintain the specified temperature, humidity and the time of mixing so as to avoid setting of cement paste.

EXPERIMENT NO. 05

• **Determination of Setting Time of Cement (IS : 4031–Part 5) 1988**

Date: ________

Objective

To determine the initial and final setting times of Cement.

Apparatus

Vicat Apparatus conforming to IS : 5513–1976.

Balance of capacity 1 kg and sensitivity to 01 gram.

Gauging trowel conforming to IS : 10086–1982.

Procedure

Unless otherwise specified this test shall be conducted at a temperature 27°C ± 2°C and the relative humidity of laboratory should be 65 ± 5%.

Prepare a paste of 300 grams of cement with 0.85 times the water required to a give a paste of standard consistency IS ; 4031–Part 4, 1988.

The time of gauging in any case shall not be less than 03 minutes not more than 05 minutes and the gauging shall be completed before any sign of setting occurs.

Fill the vicat mould with this paste resting upon a non-porous plate.

Smoothen the surface of the paste, making it level with the top of the mould.

Slightly shake the mould to expel the air.

Initial Setting Time

Immediately place the test block with the non-porous resting plate, under the rod bearing the initial setting needle.

Lower the needle and quickly release allowing it to penetrate in to the mould.

In the beginning, the needle will completely pierce the mould. Repeat this procedure until the needle fails to pierce the mould for 5 + 0.5 mm as the initial setting time.

Final Setting Time

Replace the needle of the vicat apparatus by the needle with an annular ring

Lower the needle and quickly release.

Repeat the process until the annular ring makes an impression on the mould.

Record the period elapsed between the time of adding water to the cement to the time when the annular ring fails to make the impression on the mould as the final setting time.

Report

Report the initial setting time and final setting time in minutes.

Precaution

The time of gauging in any case shall not be less than 03 minutes not more than 05 minutes.

Data Sheet

- Cement brand
- Source
- Size of sample
- Date of sampling
- Date of testing

Determination No.	(1)	(2)
Normal Consistency		
Time when water is added to Cement		
Time at initial setting		
Total time for initial setting		
Time at final setting		
Total time taken for Final setting		

EXPERIMENT NO. 06

- **Determination of Compressive Strength on Cement Mortar Cubes (IS : 4031–Part 6) 1988**

Date: ________

Definition

Compressive strength is defined as the ratio of the load per unit area.

Apparatus

Vibrating machine confirming to IS : 10080–1982.

Poking rod confirming to IS : 10080–1982.

Cube moulds shall be of 7.06 cm size confirming to IS : 10080–1982.

Gauging trowel having steel blade 10 to 15 cm in length with straight edge.

Balance of capacity 10 kg and sensitivity 01 gram.

Procedure

The quantity of cement, standard sand and water required for each cube are as follows:

Cement = 200 gms

Standard Sand = 600 gms [2.36 mm to 1.18 mm – 200 gms, 1.18 mm to 600 mic. – 200 gms, 600 mic. to 300 mic. – 200 gms]

Water = (P/4 + 3) Percentage of combined mass of Cement and Sand where P is the Consistency of Cement.

Mix dry cement and standard sand with a trowel within one minute then with required water until the mixture is of uniform colour.

The time of mixing shall be within 3 to 4 minutes.

A coating of petroleum jelly inside the moulds between the contact surface of the moulds to be applied in order to ensure that no water escapes during vibration.

Place the assembled mould on the table of the vibration machine and hold it firmly in position for two minutes for compaction of the mortar.

Place the remaining mortar in the cube mould and similarly vibrate the table for further two minutes.

Remove the mould from the vibrating machine and cut of the excess mortar with a straight edge.

Store the test specimens in a place free from vibration for 24 hours.

After this period, remove the specimens from the moulds.

Immediately submerge the cubes in a clean, fresh water or saturated lime solution and keep there until taken out just prior to test.

Conduct testing at recognized ages of the specimens, the most usual being 7 and 28 days.

Test at least three specimens preferably from different batches.

Calculations

$$\text{Compressive strength} = \frac{\text{Load}}{\text{Cross-sectional area of the specimen}}\ \text{N/mm}^2$$

The mean results of the specimens is the requirement of the Compressive strength in N/mm^2. The time of mixing is very important and in no case shall not be less than 03 minutes and not to exceed 04 minutes.

TESTS ON AGGREGATE

EXPERIMENT NO. 07

- **Determination of Specific Gravity and Water Absorption of Aggregate (IS: 2386–Part 3)**

Date: ________

Object

To determine the specific gravity of Aggregate using a container for filling water and suspending the wire basket. This test helps in the identification of stone.

Water absorption gives an idea of strength of aggregate. Aggregates having more water absorption are more porous in nature and are generally considered unsuitable unless they are found to be acceptable based on strength, impact and hardness tests.

Apparatus

A wire basket of not more than 4.75 mm mesh with hangers for suspending it from balance.

A thermostatically controlled oven to maintain temperature of 100°C to 110°C

A container for filling water and suspending the wire basket.

An airtight container of capacity similar to that of the basket.

A Balance of capacity 5 kg and sensitivity 0.5 gram.

A shallow tray and two dry absorbent cloths each about 750 × 450 mm in size.

Procedure

About 2 kg of the aggregate sample is washed thoroughly to remove fines, drained and then placed in the wire basket and immersed in distilled water with a cover of at least 50 mm of water above the top of the basket. Immediately after immersion the entrapped air is removed from the sample by lifting the basket containing it 25 mm above the base of the tank for some times. The basket and the aggregate should remain completely immersed in water for a period of 24 hours afterwards.

The basket and the sample are then weighed (W_1) g while suspended in water. The basket and the basket and the aggregate are then removed from water and are transferred the aggregate to one of the dry absorbent cloths. The empty basket (W_2) g is weighted in water.

The aggregates placed on the absorbent clothes are surface dried till no further moisture could be removed by this cloth. Then the aggregates are transferred to the second dry cloth and allowed to dry for at least 10 to 60 minutes. The surface dried aggregate is then weighed

(W_3) g. The aggregate is placed in a shallow tray and kept in an oven maintained at a temperature of 100°C to 110°C for 24 hours. It is then removed from oven and weighed (W_4) g.

Calculations

Weight of saturated aggregate suspended in water with the basket = W_1 g

Weight of basket suspended in water = W_2 g

Weight of saturated aggregate in water = $(W_1 - W_2) = W_S$ g

Weight of saturated surface dry aggregate in air = W_4 g

Weight of water equal to the volume of the aggregate = $(W_3 - W_S)$ g

(*i*) Specific gravity $= \dfrac{\text{Dry weight of aggregate}}{\text{Weight of equal volume of water}}$

$$= \frac{W_4}{W_3 - W_S} = \frac{W_4}{W_3 - (W_1 - W_2)}$$

(*ii*) Apparent Sp. Gr. $= \dfrac{\text{Dry weight of aggregate}}{\text{Weight of equal volume of water excluding air voids in aggregate}}$

$$= \frac{W_4}{W_4 - W_S} = \frac{W_4}{W_4 - (W_1 - W_2)}$$

(*iii*) Water absorption = Percentage by weight of water absorbed in terms of oven dried Weight of aggregates.

$$= \frac{(W_3 - W_4)}{W_4} \times 100$$

Limits

- The Specific Gravity of aggregate ranges from 2.5 to 3.0.
- The Water absorption of aggregates ranges from 0.1 to 2.0%.

EXPERIMENT NO. 08

- **Determination of Bulkage and Silt Content of Fine Aggregate (IS : 2386–Part 3)**

Date: ________

Introduction

Sand may contain an amount of moisture, which will cause it, when loosely filled into a container, to occupy a large volume than it would occupy if dry. It will be necessary to increase the volume of sand by the 'percentage' bulking. The correction to be made is only a rough approximation because the system of measurement by loose volume is a rough method at the best and it should be applied in order to keep the concrete uniform.

Object

This method of test covers the field method for determining the bulking and silt content of fine aggregate.

Apparatus

250 ml measuring cylinder, tray and water cane, etc.

Procedure

In the 250 ml measuring cylinder, pour the damp sand (consolidated by shaking) until it reaches the 200 ml mark. Then fill the cylinder with water and stir the sand well (the water shall be sufficient to submerge the sand completely). It will be seen that the sand surface is now below its original level. Suppose the surface is at the mark 'Y' ml the percentage of bulking of the sand due to moisture shall be calculated from the formula.

$$\text{Percentage bulking} = \left(\frac{200}{Y} - 1\right) \times 100$$

$$\text{Silt content} = \left(\frac{X - Y}{Y}\right) \times 100$$

where, X = the level of top surface of material *i.e.*, Sand + Sedimentation of silt.
Y = the level of top surface of sand layer.

Report

The percentage bulking of the sand to the nearest whole number.

Limits

Bulking and silt content is allowed maximum 10%.

EXPERIMENT NO. 09

- **Tests on Coarse and Fine Aggregates**
  (Fineness Modulus, Density and Voids Ratio of Aggregates)

Date: ________

Purpose

To obtain the fineness modulus, density and voids ratio of fine and coarse aggregate samples.

Materials and Apparatus

- Sample of fine and coarse aggregates
- Digital weighing machine
- Sieve for fine and coarse aggregates
- Various cleaning brushes

Procedure

Part–1: Sieve Analysis of fine aggregate

The process or operation of dividing a sample of aggregate into various fractions of same particle size is known as a sieve analysis. The aggregate used for making concrete are normally of size 63 mm, 50 mm, 40 mm, 25 mm, 20 mm, 16 mm, 12.5 mm, 10 mm, 4.75 mm, 2.36 mm, 1.18 mm, 600 microns, 300 microns and 150 microns. The aggregate fraction from 63 mm to 4.75 mm are termed as coarse aggregate and those fraction from 4.75 mm to 150 microns are termed as fine aggregate. The size 4.75 mm is a common fraction appearing both in coarse aggregate and fine aggregate (C.A & F.A.). Sieving can be done either manually or mechanically.

Determine the weight of aggregates that are retained in each of the sieves, by weighing each of the sieves (along with the retained aggregates), and subtracting the weight of each sieve. Also record all the weights of aggregates retained in each of the sieves. To ensure that all materials are collected, clean each sieve carefully using the proper type of brush.

Sample Calculations for Fineness Modulus, Bulk Density and Voids Ratio:

Sieve analysis of a 307 g sample of fine aggregates resulted in the following data. Find the fineness modulus.

Sieve size	10 mm	4.75 mm	2.36 mm	1.18 mm	600 μ m	300 μ m	150 μ m	< 150 μ m
Weight (g) retained	0	6	31	30	59	107	53	21

Calculations

Sieve size (1)	Mass retained in gm (2)	Percentage retained (3)	Cumulative percentage passing (4)	Cumulative percentage retained (5)
10 mm	0	0	100	0
4.75 mm	6	2	98	2
2.36 mm	31	10.1	88	12
1.18 mm	30	9.8	78	22

600 μ m	59	19.2	59	41
300 μ m	107	34.9	24	76
150 μ m	53	17.3	7	93
< 150 μ m	21	6.8	—	—
	Total = 307			Total = 246

Fineness modulus = 246/100 = 2.46

Sieve analysis of a 5 kg sample of Coarse aggregates resulted in the following data. Find the fineness modulus.

Sieve size	50 mm	40 mm	25 mm	20 mm	12.5 mm	4.75 mm	2.36 mm	600 μ m
Weight (g) retained	0	0.3	1.7	2.1	0.8	0.1	0	0

Calculations

Sieve size (1)	Mass retained in kg (2)	Percentage retained (3)	Cumulative percentage passing (4)	Cumulative percentage retained (5)
50 mm	0.0	0	100	0
40 mm	0.3	0.60	0.94	6
25 mm	1.7	0.34	0.60	40
20 mm	2.1	0.42	0.18	82
12.5 mm	0.8	0.16	0.02	98
4.75 mm	0.1	0.02	—	100
2.36 mm	—	—	—	100
600 μ m	—	—	—	100
300 μ m	—	—	—	100
150 μ m	—	—	—	100
	Total = 5.0			Total = 726

Fineness modulus = 726/110 = 7.26

• Determination of Bulk Density and Voids as per IS : 2386 (Part III)–1963

Object

This method of test covers the procedure for determining unit weight or bulk density and void of aggregates. The bulk density is the weight of material in a given volume.

Apparatus

The apparatus shall consist of the following.

Balance—A Balance of capacity 5 kg and sensitivity 0.5 gram.

Cylindrical Metal Measure—The measure shall be of 3, 15 or 30 litres capacity, according to the max. nominal size of the coarsest particles of aggregate given in Table 1.

Table 1. Size of container for Bulk Density test

Size of largest particles	Nominal capacity (litre)	Inside diameter (cm)	Inside height (cm)	Thickness of metal (mm)
4.75 mm and under	3	15	17	3.15
Over 4.75 mm to 40 mm	15	25	30	4.00
Over 40 mm	30	35	31	5.00

Tamping Rod—A straight metal tamping rod of cylindrical cross-section 16 mm in diameter and 60 cm long.

Calculation of Voids

The percentage of voids shall be calculated as follows:

$$\text{Percentage of voids} = \frac{G_S - \gamma}{G_S} \times 100$$

where, GS = Specific gravity of the aggregate and

γ = Bulk density in kg/litre

Reporting of Results

The bulk density shall be reported in kg/litre or kg/cum to the nearest 0.01 kg. The voids shall be reported as a percentage to the nearest whole number.

The condition of aggregate at the time of test shall be stated, that is (*a*) oven dry, (*b*) saturated and surface dry, or (*c*) with a given percentage of moisture.

• Buckling of Fine Aggregate as per IS : 2386 (Part III)–1963

General

It is seen that it is necessary to increase the volume of sand by the 'percentage' bulking.

Procedure

Put sufficient quantity of the sand loosely into a container until it is about two-thirds full. Level off the top of the sand. Measure the height. **Suppose this is *h* cm.**

Empty the sand out of the container into another container where none of it will be lost. Half fill the first container with water. Then add the remainder of the sand and rod it in the same way. Smooth and level the top surface of the sand and measure its depth. **Suppose this is *h′* cm.**

The percentage of bulking of the sand due to moisture shall be calculated from the formula:

$$\text{Percentage bulking} = \left(\frac{h}{h'} - 1\right) \times 100$$

Some important **bulk density** of building materials are given below:

Stone Chips

- 20 mm size = 1589 kg/cum
- 12.5 mm size = 1569 kg/cum
- Sand = 1676 kg/cum
- 1 bag of Cement = 0.0347 cum = 50 kg

EXPERIMENT NO. 10

- **Deleterious Content of Aggregates as per IS : 2386 (Part II)–1963**

Date: ________

Scope

This standard covers the following tests for aggregate for concrete:

1. Determination of clay lumps
2. Determination of clay, fine silt and fine dust (Sedimentation method)
3. Determination of light-weight pieces (coal and lignite)
4. Determination of soft particles
5. Estimation of organic impurities.

1. Determination of Clay Lumps

Object

This method of test covers the procedure for the approximate determination of clay lumps in the routine examination of aggregate. The apparatus shall consist of the following.

Balance, Containers of a size and shape that will permit the spreading of the sample on the bottom in a thin layer and Sieves.

Any particles which can be broken into finely divided particles with the fingers shall be classified as clay lumps by the use of sieves indicated below.

Size of particles making up the sample	Size of sieve for sieving residue of clay lumps
Fine aggregate (retained on 1.18 mm IS Sieve)	850 micron
Over 4.75 mm to 10 mm	2.36 mm
Over 10 mm to 20 mm	4.75 mm
Over 20 mm to 40 mm	4.75 mm
Over 40 mm	4.75 mm

Calculation

The percentage of clay lumps shall be calculated to the nearest 0.1 per cent in accordance with the following formula.

$$L = \frac{W - R}{W} \times 100$$

where, L = percentage of clay lumps,

W = weight of sample, and

R = weight of sample after removal of clay lumps.

2. Determination of Clay, Fine silt and Fine dust (Sedimentation Method)

Object

This is a gravimetric method for determining the clay, fine silt and fine dust up to 20 micron.

Test Sample

The sample for test shall be prepared from the main sample taking particular care that the test sample contains a correct proportion of the finer material. The amount of sample taken for test shall be in accordance with Table 1.

Table 1. Weight of sample for determination of Clay, Fine silt and Fine dust

Max. size present in substantial proportions (mm)	Approx. weight of sample for test (kg)
63 to 25	6
20 to 12.5	1
10 to 6.3	0.5
4.75 or smaller	0.3

A sedimentation pipette of the Andreason type of approximately 25 ml capacity and of the general form indicated in Fig. 1. This consists mainly of a pipette fitted at the top with a two-way tap and held rigidly in a clamp which can be raised or lowered as required and which is fitted with a scale from which the changes in height of the pipette can be read.

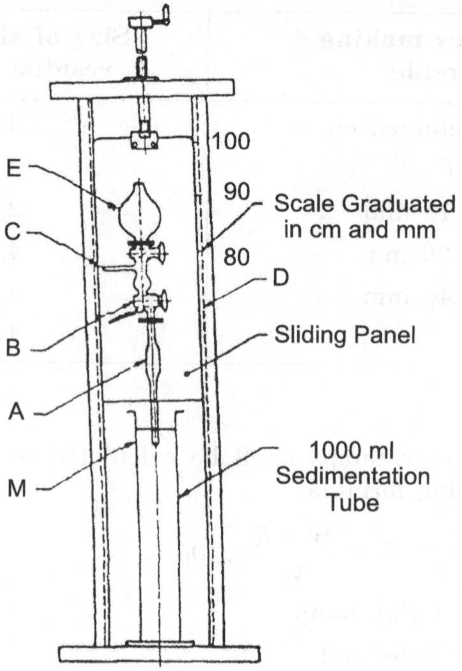

Fig. 1. Sedimentation Pipette for Determination of Clay and Silt Content

Calculations

The proportion of fine silt and clay or fine dust shall then be calculated from the following formula.

$$\text{Percentage of clay and fine silt or fine dust} = \frac{100}{W_1}\left(\frac{1000\,W_2}{V} - 0.8\right)$$

where,

W_1 = weight in g of the original sample,
W_2 = weight in g of the dried residue,
V = volume in ml of the pipette, and
0.8 = weight in g of sodium oxalate in one litre of the diluted solution.

3. Determination of Light-weight Pieces (Coal and Lignite)

Object

This method of test covers the procedure for determination of approximate percentage of light-weight pieces in aggregate by means of sink-float separation in a heavy liquid of suitable specific gravity.

The heavy liquid shall consist of a mixture of carbon tetrachloride.

Calculation

Calculate the percentage of light-weight pieces (pieces floating on the heavy liquid) as follows.

For fine aggregate:

$$L = \frac{W_1}{W_2} \times 100$$

For coarse aggregate:

$$L = \frac{W_1}{W_3} \times 100$$

where,

L = percentage of light-weight pieces,
W_1 = dry weight in g decanted pieces,
W_2 = dry weight in g of portion of sample coarser than 300 micron IS Sieve, and
W_3 = dry weight in g of portion of sample coarser than 4.75 mm IS Sieve.

4. Determination of Soft Particles

Object

This method of test deals with the procedure of determining the quantity of soft particles in coarse aggregates on the basis of scratch-hardness. This method is used to identify materials that are soft.

Calculation and Report

This report shall include the following information.

1. Weight and number of particles of each size of each sample tested with the brass rod.
2. Weight and number of particles of each size of each sample classified as soft in the test.
3. Percentage of test sample classified as soft by weight and by number of particles and
4. Weighted average percentage of soft particles calculated from percentage in item (3) and based on the grading of sample of aggregate received for examination. In these calculations, sizes finer than the 10 mm IS Sieve shall not be included.

5. Estimation of Organic Impurities

Object

This method of test covers an approximate method of estimating whether organic compounds are present in natural sand in sufficient quantities to be harmful and hence is intended to show whether further tests are necessary or desirable.

The test for determination of effect of organic impurities on mortar strength is covered by IS : 2386 (Part VI)–1963.

Harmless organic materials may cause colouration and certain naturally occurring organic compounds do not cause colouration.

TESTS ON CONCRETE

EXPERIMENT NO. 11

- **Slump Test as per IS : 1199–1959**

Date: ________

Object

This method of test specifies the procedure to be adopted either in the laboratory or during the progress of work in the field for determining, by the Slump test, the Consistency of Concrete.

The slump measured shall be recorded in terms of millimeters of subsidence of the specimen during the test. Any slump specimen which collapse or shears off laterally gives incorrect result and if this occurs the test shall be repeated with another sample.

The Procedure and Result are noted at page No. 29 and 30.

EXPERIMENT NO. 12

- **Compacting Factor Test**

Date: ________

Aim

To measure the workability of Concrete by Compacting factor test. As shown in Fig. 4.2 at page No. 31, the degree of compaction is called the compacting factor.

Apparatus

Compacting factor test apparatus.

Procedure

The sample of concrete to be tested is placed in the upper hopper up to the brim. The trap-door is opened so that the concrete falls into the lower hopper.

Then the trap-door of the lower hopper is opened and the concrete is allowed to fall in to the cylinder.

The excess concrete remaining above the top level of the cylinder is then cut off with the help of plane blades.

The outside of the cylinder is wiped clean. The concrete is filled up exactly up to the top level of the cylinder.

It is weighted to the nearest 10 gm. This weight is known as 'weight of partially compacting concrete'.

The cylinder is emptied and then refilled with the concrete from the same sample in layers approximately 5 cm deep. The layers are heavily rammed or preferably vibrated so as to obtain full compaction. The top surface of the fully compacted concrete is then carefully struck off level with the top of the cylinder and weighed to the nearest 10 gm. This weight is known as 'weight of fully compacted concrete'.

$$\text{The Compacting factor} = \frac{\text{Weight of partially compacted concrete}}{\text{Weight of fully compacted concrete}}$$

Result

The compacting factor of the given sample of Concrete is ______ %.

Observation and Calculation

Mass of cylinder = W_1

Sl. No.	Water cement ratio	Mass with partially compacted concrete (W_2)	Mass with fully compacted concrete (W_3)	Mass with partially compacted concrete only ($W_2 - W_1$)	Mass with fully compacted concrete only ($W_3 - W_1$)	Compacting factor = $\frac{(W_2 - W_1)}{(W_3 - W_1)}$
1.						
2.						
3.						

EXPERIMENT NO. 13

- **Vee-Bee Consistometer Test**

Date: ________

Aim

To measure the workability of Concrete by Vee-Bee consistometer test.

Apparatus

Vee-Bee consistometer test apparatus.

Procedure

Placing the slump cone inside the sheet metal cylindrical pot of the consistometer.

The glass disc attached to the swivel arm is turned and placed on the top of the concrete pot.

The electrical vibrator is switched on and simultaneously a stop watch is started.

The vibration is continued till such a time as the conical shape of the concrete disappears and the concrete assumes cylindrical shape.

Immediately when the concrete fully assumes a cylindrical shape, the stop watch is switched off. The time required for the shape of concrete to change from slump cone shape to cylindrical shape in seconds is known as Vee-Bee degree.

Observation and Calculation

Initial reading on the graduated rod (a)			
Final reading on the graduated rod (b)			
Slump = (b) – (a) mm			
Time for complete remolding, Seconds			

Result

The consistency of the Concrete is ________ sec.

A = Cylindrical pot.

B = Sheet metal cone

C = glass disc

D = swivel arm

E = glass disc adjustable screw

F = adjustable screw

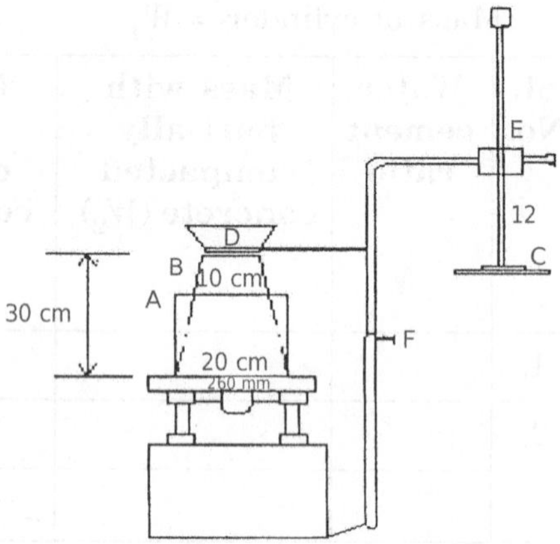

EXPERIMENT NO. 14

• **Compressive Strength of Cement Concrete Test**

Date: ________

Aim

To determine the cube strength of the Concrete of given properties.

Apparatus

Moulds for the test cubes, tamping rods, compression testing machine.

Procedure

Calculate the material required for preparing the concrete of given proportions.

Mix them thoroughly in mechanical mixer until uniform colour of concrete is obtained.

Pour concrete in the oiled with a medium viscosity oil. Fill concrete is cube moulds in two layers each of approximately 75 mm and ramming each layer with 35 blows evenly distributed over the surface of layer.

Struck off concrete flush with the top of the moulds.

Immediately after being made, they should be covered with wet mats.

Specimens are removed from the moulds after 24 hours and cured in water 28 days. Compression tests of cube specimens are made as soon as practicable after removal from curing pit. Test-specimen during the period of their removal from the curing pit and till testing are kept moist by a wet blanket covering and tested in a moist condition.

Place the specimen centrally on the location marks of the compression testing machine and load is applied continuously, uniformly and without shock. Also note the type of failure and appearance cracks.

Observation and Calculation

Specimen	Trials			Mean Value in N/mm^2
	1	2	3	
Load on cubes in kN				

Result

The compressive strength of the Cement Concrete is ________ N/mm^2.

EXPERIMENT NO. 15

- **Flexural Strength of Cement Concrete Test as per IS : 516–1959**

Date: ________

Object

Determination of the flexural strength of Concrete specimen.

Apparatus

Standard Moulds of size 15 × 15 × 70 cm for preparing the specimen, tamping rods, testing machine.

Procedure

The dimensions of each specimen shall be noted before testing. No preparation of the surface is required. The specimen shall then be placed in the machine in such a manner that the load shall be applied to the upper most surface as cast in the mould. The axis of the specimen shall be carefully aligned with the axis of the loading device. The load shall be applied with shock and increasing continuously at a rate such that the extreme fiber stress increases at approximately 7 $kg/cm^2/mm$ for the 10 cm specimens, the load shall be increased until the specimen falls. The max. load applied to the specimen during the test shall be recorded.

Calculation

The flexural strength of the specimen shall be expressed as the modules of rapture '*fb*' which if '*a*' equals the distance between the line of fracture and the nearer support measured on the center line of the tensile side of the specimen in cm shall be calculated to the nearest 0.5 kg/cm^2 as follows.

$$fb = (p \times l)/(b \times d^2)$$

When '*a*' is greater than 20 cm for 15 cm specimen or greater than 13.30 cm for a 10 cm specimen, or

$$fb = (3p \times a)/(b \times d^2)$$

When '*a*' is less than 20 cm but greater than 17 cm for 15 cm specimen or less than 13.30 cm but greater than 11 cm for a 10 cm specimen, where

b = measured width in cm of the specimen.

d = measured depth in cm of the specimen at the point of failure.

l = length in cm of the span on which the specimen was supported and

p = max. load in kg applied to the specimen.

If '*a*' is less than 17 cm for a 15 cm specimen or less than 11 cm for a 10 cm specimen, the result of the test shall be discarded.

EXPERIMENT NO. 16

- **Split Tensile Strength of Cement Concrete Test as per IS : 516–1959**

Date: ________

Object

This method covers the determination of the splitting tensile strength of cylindrical Concrete specimens.

Apparatus

Weights and weighing device, tamping rods, testing machine.

Procedure

Prepare three cylindrical concrete specimens. After molding and curing the specimens for seven days in water, they can be tested.

The bearing strips are placed between the specimen and both upper and lower bearing blocks of the testing machine.

Draw diametric lines an each end of the specimen using a suitable device that will ensure that they are in the same axial plane. Center one of the plywood strips along the center of the lower bearing block.

Place the specimen on the plywood strip and align so that the lines marked on the ends of the specimen are vertical and centered over the plywood strip.

Place a second plywood strip lengthwise on the cylinder, centered on the lines marked on the ends of the cylinder.

Apply the load continuously and without shock, at a constant rate within, the range of 689 to 1380 kPa/min splitting tensile stress until failure of the specimen.

Calculation

Calculate the splitting tensile strength of the specimen as follows.

$$T = 2P/\pi Ld$$

where, T = splitting tensile strength, kPa

P = Max. applied load indicated by testing machine, kN

L = Length, m

d = diameter, m

EXPERIMENT NO. 17

- **Rebound Hammer Test of Hardened Concrete as per BS 1881 (Part 202)–1986**

Date: ________

Object

This test is also known as the Schmidt hammer or impact hammer. It is a non-destructive method of testing concrete. The test is based on the principle that the rebound of an elastic mass depends on the hardness of the surface against which the mass impinges.

Apparatus

Rebound hammer.

Procedure

Concrete members to be tested shall be at least 100 mm thick and fixed within a structure. Dry concretes give higher rebound numbers than wet concrete. A test area shall be at least 150 mm in diameter.

Firmly hold the instrument in a position that allows the plunger to strike perpendicularly to the surface tested. Gradually increase the pressure on the plunger until the hammer impacts.

After impact, record the rebound number to two significant figures. Take ten readings from each test area. No two impact tests shall be closer together than 25 mm.

Calculation

Discard readings differing from the average of 10 readings by more than 5 units and determine the average of the remaining readings. If more than 2 readings differ from the average by 7 units, discard the entire set of readings.

Some points of maintenance of buildings/structures are necessary for better performance as it enables the buildings to safely complete its full life span without any further deterioration as noted below:

Structurally Distressed RCC Members of Buildings

Buildings and other concrete structures suffer deterioration due to a variety of reasons. The treatment should be limited to locations where the problem occurs on the effected part of the body. Based on the efficiency of various techniques like, jacketing, etc. in restoration can be considerably enhanced it, this approach is incorporated in the methodology adopted.

Guniting: Effective Technique

Guniting is an effective technique which has been extensively used in the restoration of structurally distressed RCC members. The present practices wherein cover concrete is removed on a mass scale and every good/sound concrete is removed forcibly by chisel and sledge hammer, has to make way for a more focused treatment of the affected part of the structure. It is true that as the old concrete surface is coated with bond coat before placing polymer mortar, there is reason to believe that the restored portion takes the load as it did prior to the restoration.

The guniting technique suffers from some drawbacks like dust and noise nuisance. Though the guniting technique is now being increasingly replaced by other methods, the following points need to be kept in mind for better results:

- Coating of existing as well as new bars by zinc rich epoxy primer to guard against corrosion.
- Mesh reinforcement is not advised. It is advised reinforcement only if it is required for structural purposes.

Restoration by Polymer Mortar/Micro Concrete

As of today, this methodology appears to be the best one for restoration. It is absolutely necessary to treat the structurally distressed RCC members both externally and also from the inside of each and every flat.

One such example of structural deterioration of building was noticed at IICT, Hydrabad. The condition of main laboratory building was deteriorated due to various reasons. The main reason for structural deterioration of building was due to leakages of chemicals flowing through channels made inside the building at all the floors. The continuous flow of chemical/ waste materials through these drains ingressed through the surface of RCC floors, beams and slabs resulting in carbonation of concrete and corrosion of reinforcement steel. The said building got structurally repaired by micro concreting after treatment of corrugated steel reinforcement bars with anti corrosive drainage pipes were relocated outside the building and the building is being renovated completely.

Jacketing

In this method the entire height of the column section is increased and a cage of additional main reinforcement bars with shear stirrups is provided right from the foundation as per the requirement of additional load, etc.

INDEX